Elements of Optical Networking

Volkmar Brückner

Elements of Optical Networking

Basics and Practice of Glass Fiber Optical
Data Communication

Second Edition

 Springer Vieweg

Volkmar Brückner
Werl, Germany

ISBN 978-3-658-43240-9   ISBN 978-3-658-43242-3   (eBook)
https://doi.org/10.1007/978-3-658-43242-3

This Springer Vieweg imprint is published by the registered company Springer Fachmedien Wiesbaden GmbH, part of Springer Nature.
The registered company address is: Abraham-Lincoln-Str. 46, 65189 Wiesbaden, Germany

Paper in this product is recyclable.

Dedicated to my teacher and friend Prof. Dr. Bernd Wilhelmi (†2018), from whom I was able to learn how to deal with science and people

Foreword to the Second Edition

About 12 years have passed since the first edition of the textbook "Elements of Optical Networks" was published in English, years that were marked by many events such as Corona, cyber–attacks, Russia's war against Ukraine etc. What has always remained and grown even more is the need for communication. Mobility in communication (keyword: smartphones as little all–rounders) has increased noticeably, more than 8 billion mobile phones worldwide - more than the world's population - are proof of this. Even if the distribution worldwide continues to be a problem, we have moved a little closer to anytime, anywhere communication. The development of increasingly versatile forms of mobile communication in global networks has been and still is in the foreground, but the importance of fibre optic networks as the basis of communication has continued to grow.

The optical transmission of data and information via optical fibres is on the rise worldwide - the laying of optical fibres and entire optical fibre networks as well as the construction and further development of corresponding lasers, receivers, amplifiers, couplers, and switches continues unabated.

Therefore, the demand for well–trained specialists at university level and the continuous training of employees will be and remain of enormous importance for the operators of optical transmission systems. The Second edition of my book "Optical Communications Engineering - Fundamentals and Applications" is also dedicated to this training and further education.

Teachers in the field of optical communications technology are thus faced with the task of didactically preparing these processes, some of which are very complicated, for students at universities as well as for further education, e.g., for communication technicians. This textbook is based on basic knowledge of physics and mathematics. In doing so, I consciously make use of many years of lecture experience at the Deutsche Telekom University of Applied Sciences in Leipzig and at foreign universities. From many conversations, especially with students, it emerges again and again that a seemingly clear matter is far from being clear and that new questions and problems arise again and again.

The central concern of this textbook is therefore to present the problems in a comprehensible way and to translate them into formulas or recommendations that can be used in practice. This naturally goes hand in hand with a primarily qualitative approach, without penetrating into all the details of the formal mathematical apparatus. The description of the processes therefore comes first. This common thread is explained by many examples and sample calculations as well as by illustrating interrelationships by means of simple mathematical programmes. Furthermore, the problems or limits of a purely optical data transmission are explained by means of practical examples. After reading this book, students should be able to understand and classify the latest developments in the field of optical communications technology in the technical literature or at technical conferences.

Clear structure, detailed explanations with a minimum of mathematical effort - this has already been praised in previous editions - this style is of course maintained. The book is also available in an electronic version. Partly in contrast to the printed book, the "e–book" contains links to corresponding websites as well as links to supplementary or more detailed explanations.

In addition to the examples already given in the 1st edition, there are now numerous exercises. In the written version, the solutions are given in Chap. 12, in the electronic version, links lead alternatively to formulas (in the text as "Help") or to the solution (in the text as "Solution").

Once again, some complicated relationships are illustrated by example calculations carried out in MathCad. The corresponding programmes are included in the electronic version both as a text version and as a fully executable programme. Further MathCad programs and supplements to the book, as well as information posted later, are available for download in the publisher's OnlinePlus area.

The author offers the illustrations from this book as additional material in the Online-Plus area of the publisher. In agreement with the publisher, these pictures can be used for teaching and further education purposes - however, I ask that the source be named and, if possible, that the author be informed (v.brueckner@hotmail.de). Under the above e–mail address, I am also happy to receive suggestions and comments - a book can only be improved later if the weak points are known.

In terms of content, the book is based on the German–language edition [Brü 23]. The software from DeepL (www.deepl.com) was a great help to me in the translation.

In the meantime, the centre of my life has moved from Leipzig in Saxony to Werl in North–Rhine Westphalia. Since the everyday stress of working life has been eliminated, I have had more opportunities to reflect on the design of teaching and to test it internationally. I would like to incorporate these experiences into this book.

My wife Ute deserves thanks for allowing me the time to write this new edition - other important things have been left undone or delayed as a result.

Werl   Volkmar Brückner
January 2024

Contents

Introduction

1

The optical transmission of data and information is one of the very old techniques. Even the hunter-gatherers used smoke signals. In the twelfth century BC, the ancient Greeks are said to have "reported" the fall of Troy with coded torch signals (one torch each in the left and right hands means α, two torches on the left, one torch on the right means β, etc.). Eight characters per minute are said to have been achieved - which corresponds to a "bit rate" of 0.13 characters per second. A few centuries earlier, these techniques had already been used on the Great Wall of China - information about the approaching enemy was provided "optically". With the "wing telegraph" developed by Claude Chappé in 1794 (comparable to railway signaling systems), it was possible to transmit a message in less than an hour over about 300 km from Lille to Paris. Signalmen posted along the route observed the position of the winged telegraph with binoculars and relayed the information. The "modern heliograph" of 1875 used the reflection of light by means of mirrors, the messages were transmitted with a kind of Morse signal. Even today, the election of the Pope is indicated by white smoke—in other words, optically! However, the signal rates transmitted with these "old" optical transmission techniques - we would call them bit rates today - were very low, often less than a few bits per second.

Optical transmission from point A to point B follows a uniform pattern (Fig. 1.1).

Today, digital transmission technology dominates, whereby every piece of information, which is generally analogue, must first be digitized: For speech, the individual frequencies, and their proportions in the signal (volume) are determined and digitally encoded. An image is divided into pixels point by point, and for each pixel, the brightness, the color composition as well as the position are digitally encoded; for letters, one has an "arrangement rule" for pixels and so on. It can be seen that the information about a single pixel or a single sound already requires a larger number of digital signals - bits.

Supplementary Information The online version contains supplementary material available at https://doi.org/10.1007/978-3-658-43242-3_1.

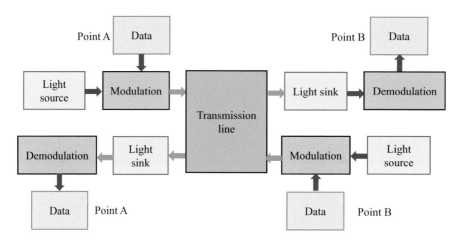

Fig. 1.1   Optical transmission between point A and point B (bidirectional)

This creates a bit sequence, the so-called bit pattern, a certain sequence of yes–no, on–off, or 1–0 information (read: one-bit or zero-bit). These seemingly lengthy processes actually run extremely fast - thanks to an extremely high transmission rate.

This leads to the necessity of transmitting huge amounts of data. To estimate the bit rates to be transmitted, consider, for example, the situation for transmitting live images from an operating theater to a specialist at a distant point on earth. For example, a single image on a computer at a high resolution of 1024×768 bits and 32-bit color depth contains the data quantity 25.2 Mbit (Mb); a good TV display is obtained at 100 frames per second. If one wants to transmit this amount of data, one will have to transmit $2.5\ 10^9$ bits per second (2.5 Gbps). Without data compression methods, the individual bit may therefore only be less than $1/2.5 \cdot 10^9 = 4 \cdot 10^{-10}$ s $= 400$ picoseconds (400 ps) "long". Such huge amounts of data can only be transmitted over long distances (i.e., over more than 1 km) in optical fibers.

Today, fiber optic networks are used to transmit high digital data rates in the range of gigabits per second over long distances. Optical transmissions in fiber optic networks must fulfill two conditions in particular:

- All operations known from electrical transmission technology must be realized optically. As far as possible, there must be no interfering intermediate steps in the non-optical range (all-optical system) and the optical system must be potentially more efficient than the electrical system.
- The optical and electrical systems must be compatible with each other, because despite all progress, both systems will exist in parallel in the long term - e.g., it is not conceivable that all house connections will be exclusively optical in the foreseeable future. In addition, compatibility with radio transmission (e.g., via mobile radio or satellite technology) must be guaranteed.

The central element of optical communications technology is the optical fiber itself, which is often constructed as a ring structure, more rarely as a bidirectional end-to-end or point-to-point connection. In the meantime, millions of kilometers of optical fiber have been laid worldwide.

For example, in 1997 the 27,000 km long Fiber-optic Link Around the Globe (FLAG) was installed through 20 countries at a cost of about €500 million along the former Silk Road. Less well known is that fiber optic cables are also laid on the ocean floor.

Another example is the "TransAtlantic Telephone Cable" (TAT). TAT-14 as a submarine cable ring connects Manasquan and Tuckerton in the USA with North/D since 2001 over a total of 15,000 km through 4 fiber pairs with a transmission capacity of 640 Gbps. One route starts in Norden and runs via Blaabjerg/DK and the Shetland Islands (Scotland) to Manasquan and Tucker-ton (New Jersey). Another route also starts in the north and leads via Katwijk/NL, St. Valery-en-Caux/F, Bude (England) through the Atlantic again to Tuckerton and Manasquan. The fiber optic cables are connected via splices and cable sleeves. There are also amplifiers on the seabed at intervals of 50–70 km, which are supplied with power via a copper cable inside the TAT fiber optic cable. Unnoticed by the public, TAT-14 was decommissioned on December 15, 2020. After nineteen years, the technology was obsolete. The operators found it cheaper to lease capacity on newer lines.

With 39,000 km, the submarine cable route South-East Asia-Middle East-Western Europe 3 (SEA-ME-WE 3) is even longer. Since 1999, it has connected 33 countries on 4 continents (Australia, Asia, Africa, and Europe) via 39 landing points.

Today, we are talking about global networks. The main element of such networks is the so-called backbone, the fiber optic network.

Figures 1.1 and 1.2 show a so-called point-to-point connection with optical fibers: A source (source 1) is digitized by an Analogue-to-Digital (AD) converter, resulting in an electrical bit sequence (Fig. 1.1). The digital signal can also be generated directly (e.g., by a computer) (source 2). This electrical bit sequence is converted by a modulated laser beam into an optical bit sequence (e → o converter), which is coupled into an optical fiber. The optical fiber can be extended by means of an optical coupler. In addition, if necessary, the optical bit sequence is amplified by an optical amplifier (Fig. 1.1). The reverse situation occurs at the end of the transmission path (Fig. 1.2). The optical, modulated signal is converted back into an electrical signal in a receiver and, after demodulation, is available as digital data directly (e.g., in a computer, sink 2) or, if necessary, after Digital-to-Analogue (DA) conversion, again for the terminal devices as analogue (sink 1). Receiver and demodulator are therefore also components of an optical network.

This names essential elements of an optical network that are described in this book: Optical fiber with attenuation, dispersion and nonlinearities, connection of optical fibers by connectors and splices, semiconductor lasers and modulators, and optical amplifiers. Important is the generation of an "optical data stream". The central element here are semiconductor lasers that operate at a (selectable) wavelength λ_1 in the range around 1550 nm with an extremely small line width $\Delta\lambda$ of less than 10^{-4} nm. Let's call the

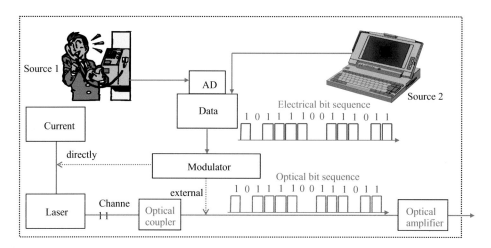

Fig. 1.2 Modulated light source (source)

semiconductor laser (light) channel number 1. The source of the data can be very different (Fig. 1.1), e.g., a normal telephone - then you need an analogue-to-digital converter
(AD) - or digital electrical data directly from a computer. This laser light is modulated
either directly when it is generated or subsequently (externally) by means of a modulator,
which in turn is controlled by the data to be transmitted (Fig. 1.1). This power modulation creates an optical bit sequence. Altogether, this forms the light source in channel 1
(Fig. 1.1), which can then be "transported" over long distances (many thousands of kilometers) via a fiber optic network (Fig. 1.3).

Two or more light sources can also be transmitted with one (common) optical fiber—
this is done by (wavelength) multiplexing (MUX, Fig. 1.4a). At the end of the transmission path, the signals must then be separated again by means of a demultiplexer
(DEMUX, Fig. 1.4b). The separation of very closely spaced channels with ≥ 0.3 nm
channel spacing is a particular challenge for engineers. These techniques will also be
presented in the book.

The optical bit sequences prepared in this way are coupled into the optical fiber ring
(add) or, conversely, "taken out" of it (drop) with an optical add-drop multiplexer. The
aim is to achieve an optimal distribution of the data streams over the individual optical
fibers or wavelengths.

This may require changing the channel, which is realized by optical cross-connectors. If both elements are realized purely optically (i.e., without intermediate conversion
to the electrical domain), they are referred to as optical add-drop multiplexers (OADM)
or optical cross-connectors (OCC or OXC), where the "X" is intended to symbolize the
crossing of optical lines (Fig. 1.5). In the OADM or OCC, (passive) couplers and (active)
switches must be present - reason enough to describe how they work.

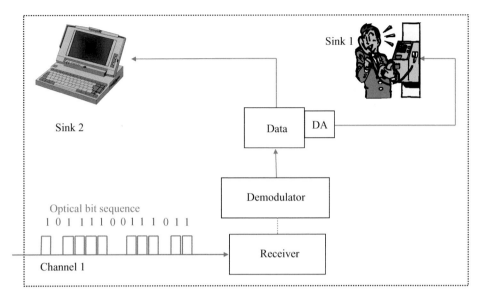

Fig. 1.3 Receiver area (sink)

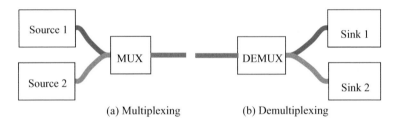

Fig. 1.4 Multiplexing and demultiplexing

It is not far from the fiber optic skeleton to the global network with long-distance connections. Other networks are grouped around the fiber optic network: The (mostly already existing) fixed network, in which the lines consist of copper wires (mostly they are part of the "old" telephone network), the mobile network, in which the transmission is done by electromagnetic waves, e.g., to the smartphone, and the satellite radio network, where electromagnetic waves are sent from the transmitter to a geostationary satellite and from there back to earth to the receiver [Brü 22].

Transitions are necessary between all these networks: Between an optical local area network (oLAN) and the optical fiber network, between mobile radio wave networks and the optical fiber network, between electrical networks and the optical fiber network, between satellite radio wave networks and the optical fiber network, between satellite radio wave networks and the optical fiber network, between wireless local radio net-

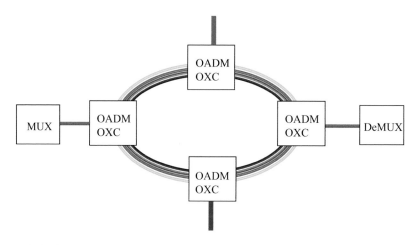

Fig. 1.5 Optical add-drop multiplexers (OADM) and cross-connectors (OXC)

works and electrical networks, etc. This book will also describe the access points to and from the optical network.

Optical fibers dominate in optical networks - complementary to this, polymer optical fibers play an increasing role in the short range [Zie 08]. However, since the basic approaches and problems are similar, only optical fibers will be discussed here.

The aim of this book is to present the partly complicated interrelationships in a simple and understandable way, to describe the functioning of the individual elements with a view to practical applications, and to show difficulties in commercial realization and problems in the interaction of the individual elements in an optical network by means of examples and tasks.

The content of the book is complemented by exercises, for example:

Exercise 1.1:

Practice the handling of tasks!

The solutions to the tasks can be found in the printed book as an appendix. In the eBook, there is a link from the electronic text to help and solutions, for example, are the help and the solution of the above task (Exercise 1.1):

▶ **Tip**
   Help H1-1 (Sect. 12.1).
   Solution S1-1 (Sect. 12.2).

In the electronic version of the book, you can click on the "Help" and "Solution" functions via links. Furthermore, you will find additional texts and mathematical programs in the folder "Extras" via a link provided by the publisher. It is recommended to activate the link initially and to keep it open while reading the book. The additional materials as well as the mathematical programs in MathCad are linked directly from the text to the "Extras" folder. The corresponding text version is also available in a pdf file. The same applies to links from the text to corresponding websites.

The partly complicated relationships are illustrated by example calculations, which were carried out in MathCad. MathCad allows a flexible representation of correlations and dependencies of parameters. Mathcad can be downloaded free of charge (for students and non-professionals) from the internet (in German or English), for example, https://www.google.rw/search?q=mathcad+15+free+download&ie=&oe=. In the electronic version, the corresponding basics are available, e.g., for the representation of a parabola in the "Extras" folder as a program link. A corresponding text version as a pdf file (also in the "Extras" folder) is of course also available as a link. The example also gives some hints for working with MathCad.

Many additional materials, which should be quite helpful for concrete problem solutions, as well as the corresponding executable mathematical programs in MathCad are also available for download—in the printed book via the Springer Verlag website as "Online additional materials" in the "Extras" folder.

A comprehensive theoretical description of the processes in glass fibers can be found, for example, by Saleh [Sal 08]. Regardless of this, it is sometimes necessary to go into theoretical basics for understanding (e.g., for the wave image and interference).

There is, of course, a great deal of literature on this subject. I have only cited those sources that are close to my understanding of the problem or from which I have taken necessary information on problems that I could not or did not want to go into detail. I have largely avoided literature on specific technical papers—access to this literature can best be found from the reports of annual technical conferences (e.g., European Conference on Optical Communication, ECOC).

This book should enable the reader to better master studies and to understand the latest developments in the field of optical data transmission and, if necessary, to apply and expand them in engineering terms.

Light

2

2.1 What is light?

Light is - similar to radio waves - electromagnetic radiation. Light is characterized by

- Power (measured in watts) or energy (for light pulses, measured in Watt-seconds Ws = Joule J). For practical tasks, the power or energy density as power or energy per area is also important. Details can be found under Radiation Units.
- The spectrum of light. The spectrum is the dependence of the light intensity on the wavelength or the frequency f. The relationship between the two is as follows. There is a relationship between frequency and wavelength:

$$f \cdot \lambda = c_0 \tag{2.1}$$

with the speed of light in a vacuum $c_0 = 2.99792458 \cdot 10^8 \, \mathrm{m/s}$.

If we consider a real medium with refractive index n, Eq. (**2.1**) will be modified:

$$f \cdot \lambda = \frac{c_0}{n} \tag{2.2}$$

In some cases, light will also be described by the *photon energy* E:

$$E = h \cdot f = \frac{h \cdot c_0}{\lambda} \tag{2.3}$$

With Planck's constant, $h = 6.6260755 \cdot 10^{-34} \, \mathrm{Ws^2} = 4.135670 \cdot 10^{-15} \, \mathrm{eVs}$. The unit of photon energy is *electron volt* (eV).

Supplementary Information The online version contains supplementary material available at https://doi.org/10.1007/978-3-658-43242-3_2.

© Springer Fachmedien Wiesbaden GmbH, part of Springer Nature 2024
V. Brückner, *Elements of Optical Networking*,
https://doi.org/10.1007/978-3-658-43242-3_2

Tab. 2.1 Spectrum of electromagnetic waves

Range	Short name	Wavelength	Frequency (Hz)	Photon energy (eV)
γ-emission	γ	<500 pm	>6·10^18	>24.8·10^3
X-ray emission	X-Ray	<50 nm	>6·10^15	>24.8
Ultraviolett	UV	<400 nm	>7.5·10^14	>3.1
Visible	VIS	<700 nm	>4.3·10^14	>1.77
Infrared	IR	<100 μm	>3·10^12	>12.4·10^{-3}
Microwaves		<1 cm	>3·10^10	>124·10^{-6}
Radio waves	RF	<1 km	>3·10^6	>12.4·10^{-9}

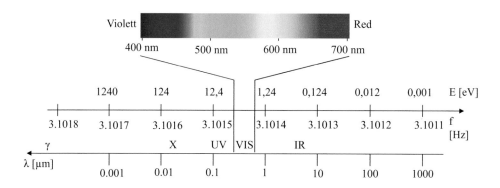

Fig. 2.1 Electromagnetic radiation and visible light (VIS)

One can also use a modification of Eq. (2.3) which contains units (E in eV, λ in μm).

$$E/eV = \frac{1.23984}{\lambda/\mu m} \qquad (2.4)$$

For example, wavelength λ = 1.5 μm corresponds to a photon energy E = 0.826 eV.

The spectrum consists of several parts (Tab. 2.1).

In photonics, "light" is the range from UV to IR. Technically speaking - and also in the colloquial - light is the visible part of the spectrum (VIS). One can see the visible part of the spectrum in a rainbow - that's why we sometimes speak of *rainbow colors* (Fig. 2.1): violet, blue, green, green-yellow, yellow, orange, and red. In communication technology, wavelengths in near-infrared (NIR) range (in glass fibers) and sometimes in VIS (plastic fibers) are used.

Exercise 2.1

Which frequency and photon energy correspond to a wavelength of 1550 nm?

▶ **Tip**
   Help H2-1 (Sect. 12.1)
   Solution S2-1 (Sect. 12.2)

2.2 Particle or Wave—Duality of Waves and Particles

In daily life, one can find many examples where light can be considered as ray, e.g., sunlight in the woods in the morning (Fig. 2.2).

Furthermore, there are numerous occurrences which can only be explained by the wave character of light, e.g., colors in an oil layer (Fig. 2.2). So: What is the right answer to the question - is light a ray or a wave or both?

In 1676, Isaac Newton (1642–1727) developed the particle model: Light consists of massless balls propagating straightforward with the speed of light. Today, these are called photons or quanta (from the Latin quantum: how big or how much) of the electromagnetic field.

The wave model was developed around 1690 by Christiaan Huygens (1629–1695). An important highlight supporting the wave picture was the development of Maxwell's equations by James Clerk Maxwell (1831–1879).

Late in the nineteenth century, the wave picture of Huygens seemed to be the better theory. After the wave theory seemed to have triumphed in the late nineteenth century and only "insignificant" questions were still open, in 1900, Max Planck (1858–1947)

Fig. 2.2 Sunlight as rays (left) and interference in oil (right)

developed the *quantum optics*. Albert Einstein (1879–1955) used this theory to explain the photoelectric effect in 1905, which earned him the Nobel Prize. He also coined the term "light quanta" (photons).

The symbolic quantum theory of light has been in development since 1925, starting with papers by Max Born (1882–1970), Ernst Pascual Jordan (1902–1980), and Werner Heisenberg (1901–1976). Up to now, the theory of electromagnetic radiation—the so-called quantum electrodynamics—describes the quantum nature of light the best.

To combine both of these pictures—the particles model and the wave model—the essential proposal was done by Louis-Victor Pierre Raymond de Broglie (1892–1987).

De Broglie postulated the dual character of light. He stated that mass-carrying particles can be characterized by wave properties: He considered two essential things in physics (on the left the mechanical quantities E and p, and on the right the wave quantities f and λ):

- Law of conservation of energy: The kinetic energy in classical mechanics E_{kin} of particles of mass m and velocity v is proportional to the frequency f or the wavelength λ of a wave:

$$E_{kin} = \frac{m}{2} \cdot v^2 = h \cdot f = \frac{h \cdot c_0}{\lambda} \qquad (2.5)$$

The particle character of light is expressed in Planck's quantum of action h.

- Law of conservation of momentum: The momentum in classical mechanics (as a vector) $\vec{p}_{mech} = m \vec{v}$ (particle) is inversely proportional to the wavelength λ of the wave:

$$p_{mech} = m \cdot v = \frac{h}{\lambda} \qquad (2.6)$$

In terms of light, the dualism of wave and particle leads to the dualism of electric and magnetic fields that propagate in waves (electromagnetic wave), which can be used to explain and calculate many properties of light (e.g., color, interference, diffraction, and polarization) and the ray pattern (geometric optics). It should be noted that De Broglie's postulate is arbitrary and has been supplemented and extended in modern physics by quantum physical states (keyword: entangled states, see also Sect. 2.2.2).

2.2.1 Wave Viewpoint

The original idea of wave propagation similar to water waves was also taken up for light. This can be used to describe many phenomena, especially the ability to superimpose or interfere. This is the *diffraction* of light.

The wave equation can be derived from Maxwell's equations; it describes the change in field strength (as a vector) $\vec{E}(\vec{r},t)$ of the electric field with respect to location (as a vector) $\vec{r}$ and time t:

$$\nabla^2 \vec{E}(\vec{r},t) = \frac{n^2}{c^2} \frac{\partial^2 \vec{E}(\vec{r},t)}{\partial t^2} \quad \text{oder} \quad \Delta \vec{E}(\vec{r},t) - \frac{n^2}{c^2} \frac{\partial^2 \vec{E}(\vec{r},t)}{\partial t^2} = 0 \quad (2.7)$$

Here, ∇ is the so-called Nabla operator and Δ the Laplace operator. Operators are calculation rules, e.g., the Nabla operator $\nabla = \left(\frac{\partial}{\partial x}, \frac{\partial}{\partial y}, \frac{\partial}{\partial z} \right)$ is a vector, the Laplace operator $\Delta = \nabla^2 = \frac{\partial^2}{\partial x^2} + \frac{\partial^2}{\partial y^2} + \frac{\partial^2}{\partial z^2}$ on the other hand is a scalar. Equation (2.7) is then written as

$$\frac{\partial^2 E(x)}{\partial x^2} + \frac{\partial^2 E(y)}{\partial y^2} + \frac{\partial^2 E(z)}{\partial z^2} = \frac{n^2}{c^2} \frac{\partial^2 \vec{E}(t)}{\partial t^2} \quad (2.8)$$

The wave equation is the basis for all calculations in wave theory. Here we are only interested in the solution of the wave Eqs. (2.7) or (2.8). Details of the calculation can be found as a pdf file in the "Extras" folder.

We choose an approach with a cos-function and consider that the wave should propagate in the z-direction (longitudinal direction), x and y are therefore the transverse directions (transversal direction):

$$E(x, y, z, t) = E_0(x, y) \cdot \cos(\phi) = E_0(x, y) \cdot \cos(\omega t - kz) \quad (2.9)$$

E_0 is the amplitude, ϕ is the phase, k is the wave number $k = n \cdot \frac{\omega}{c_0} = n \cdot k_0 = n \cdot \frac{2\pi}{\lambda_0}$, ω is the angular frequency $\omega = 2\pi \cdot f$, and c_0, λ_0, and k_0 are the speed of light, wavelength, and wave number in vacuum, respectively.

For concrete calculations (as by MathCad program, see also pdf file) it is better to use the representation of the wave in complex form instead of the approach according to Eq. (2.9).

$$E(x, y, z, t) = \frac{1}{2} E_0(x, y) \cdot \left[e^{-i(\omega t - kz)} + e^{+i(\omega t - kz)} \right] \quad (2.10)$$

It can be shown that Eqs. (2.9) and (2.10) are equivalent.

So, we have a wave that propagates in the z-direction with field strengths in the x- and y-directions. These are so-called transverse waves. Let us consider electric field strengths only in x direction, i.e., $E_y = E_z = 0$ (Fig. 2.3).

Figure 2.3 shows the location- or time-dependent part of the wave propagation. After a period of 2π, both E(z) and E(t) reach the same value. The wavelength λ is the value on the spatial axis at which E(z) has assumed the same value again after passing through a period. The period of oscillation T is the time after which E(t) has assumed the same value again. The reciprocal period of oscillation is the frequency f = 1/T. If one "sits" on a certain point of constant phase ($\omega \cdot t - k \cdot z$), e.g., on a wave crest in Fig. 2.3, one

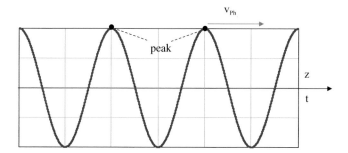

Fig. 2.3 Waves in the z-direction (space periodicity) or time direction (time periodicity)

"rides" on this wave (e.g., at peak point) crest in the z-direction with the phase velocity v_{Ph}

$$v_{Ph} = \frac{dz}{dt} = \frac{d\left(\frac{\omega}{k} \cdot t - const\right)}{dt} = \frac{\omega}{k} \tag{2.11}$$

Therefore, the distance between neighboring peak points in Fig. 2.3 is exactly λ.

Exercise 2.2

Prove that the approach from (2.9) is a solution of the wave equation.

▶ **Tip**
Help H2-2 (Sect. 12.2)
G:\Geteilte Ablagen\ST GLS Projekte\1-Frontlist_A-E\Brückner, Elements of Optical Networking\WeTransfer\Solutions.docx - S2-2

Exercise 2.3

Show that the approaches from Eqs. (2.9) and (2.10) are identical using Euler's identity $e^{\pm ix} = \cos x \pm i \cdot \sin x$!

▶ **Tip**
Help H2-3 (Sect. 12.1)
G:\Geteilte Ablagen\ST GLS Projekte\1-Frontlist_A-E\Brückner, Elements of Optical Networking\WeTransfer\Solutions.docx - S2-2

The generation of a wave can be imagined as a point source radiation of spherical waves. As a mechanical analogue, we can consider the circular water waves starting at the point of impact where a stone is thrown into the water. Looking at light propagation over a large distance from the source, i.e., under a very small radiation angle, we see only slightly bent wavefronts - i.e., plane waves (Fig. 2.4).

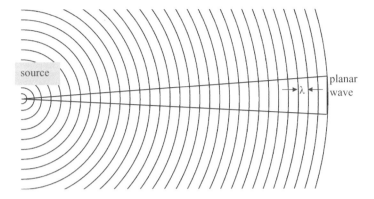

Fig. 2.4 Spherical and plane waves

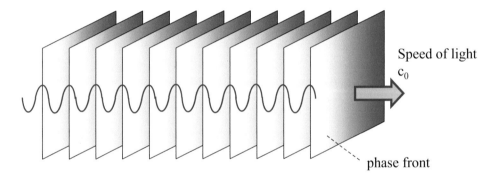

Fig. 2.5 Phase fronts of waves

One can use also phase fronts to illustrate propagation of planar waves in the z-direction with phase speed c_0 (Fig. 2.5).

2.2.2 Ray Viewpoint

The modern particle theory (quantum theory) considers photons as massless spheres of energy h·f, which rotate around their own axis. This rotation is called spin, the rotation axis is the polarization of the photons.

Now consider a pair of photons emitted from a common light source. Following the thought experiment of Albert Einstein (1879–1955), Boris Podolsky (1896–1966) and Nathan Rosen (1909–1995) from the year 1935 (EPR paradox) this photon pair has no common polarization after the creation, the polarization is "blurry". Only by a measurement or action of an optical element, e.g., a polarizer, a polarization is assigned to the photon. Now, the strange or paradoxical thing is that, similar to identical twins, the determination of one photon forcibly leads to the determination of the other photon—and this is independent of the distance between the two photons. According to the quantum mechanical representation, neither of the photons has a fixed polarization, but nevertheless, both always behave in the same way. Such pairs of photons are called entangled photons. This contradicted previous quantum mechanics. Einstein called this behavior "spooky action at a distance".

Let us now look at the propagation of photons in the z-direction. For these purposes, we can simplify Fig. 2.5. Then we get a picture like in Fig. 2.6. Now we draw only phase fronts of planar waves, the distance of phase fronts is exactly λ. In the x- and y-directions, the phase fronts are infinite. In Fig. 2.6a, the ray (arrow) with corresponding phase fronts and the cos wave are depicted. Often only the arrow or the arrow with phase fronts is used (Fig. 2.6b).

Many features in optics can be clearly described qualitatively (sometimes even quantitatively). Still, it seems necessary to keep the wave viewpoint in mind.

In particular, one can describe the *refraction* of light by ray viewpoint. The ray image fails for the description of superposition effects. One can state: The interaction of rays (e.g., diffraction and penumbra) cannot be described with the ray model. In any case, the wave image must also be kept in mind.

The ray image is often used in classical ray optics to describe the ray path in lenses or materials with different refractive indices (ABCD matrix model).

Exercise 2.4

Which kind of waves can be described by the ray model?

▶   **Tip**
Solution S2-4 (Sect. 12.2)
G:\Geteilte Ablagen\ST GLS Projekte\1-Frontlist_A-E\Brückner, Elements of Optical Networking\WeTransfer\Solutions.docx—S2-2

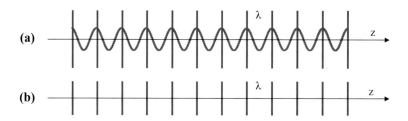

Fig. 2.6 Simplified representation of a planar wave as cosine (a) and arrow (b)

2.3   Power and Energy of Electromagnetic Waves

Electrical engineering uses the terms power (as the product of voltage and amperage, Units: V·A or Watt, W) and energy (as the product of power and time, Units: VA·s or Watt-seconds, Ws). Light as an electromagnetic wave can be similarly described by *physical power* P (Units: W or mW). In engineering sciences, the term *engineering power* p (or engineering level measure) is often used with the unit dBm - this is power in mW with respect to a reference power 1 mW on a logarithmic scale

$$p(\text{in dBm}) = 10 \cdot \lg \left(\frac{P(\text{in mW})}{1 \text{ mW}} \right) \qquad (2.12)$$

Thus power 0 dBm corresponds to 1 mW, doubling of power (2 mW) corresponds to $p = +3$ dBm, 0.5 mW corresponds to $p = -3$ dBm, etc.

It is not uncommon in engineering to refer to the level measure as power, which can be very confusing and cause problems.

Exercise 2.5

Which power p corresponds to 100 mW physical power?

▶ **Tip**
Help H2-5 (Sect. 12.1)
Solution S2-5 (Sect. 12.2)
G:\Geteilte Ablagen\ST GLS Projekte\1-Frontlist_A-E\Brückner, Elements of Optical Networking\WeTransfer\Solutions.docx—S2-2

The difference between two powers gives the *attenuation* a. It is conventionally defined in units of decibel (dB):

$$a(\text{in dB}) = p_1(\text{in dBm}) - p_0(\text{in dBm}) = -10 \cdot \lg \left(\frac{P_0(\text{in mW})}{P_1(\text{in mW})} \right) \qquad (2.13)$$

Exercise 2.6

Which attenuation results if power is reduced from 100 to 50%?

▶ **Tip**
Help H2-6 (Sect. 12.1)
Solution S2-6 (Sect. 12.2)
G:\Geteilte Ablagen\ST GLS Projekte\1-Frontlist_A-E\Brückner, Elements of Optical Networking\WeTransfer\Solutions.docx—S2-2

The power drop in the z-direction (e.g., propagation in a glass fiber) can be described by *Lambert-Beer's law* discovered by Pierre Bouguer (1698–1758) in 1729:

$$P(z) = P_0 \cdot e^{-\alpha' z} \tag{2.14}$$

where α' is an attenuation factor (unit: km^{-1}).

To describe the power drop (or as pdf file) in terms of dBm one can use the relation:

$$p_1(z) - p_0 = -4,343 \cdot \alpha' \cdot z \tag{2.15}$$

The corresponding relations are depicted in Fig. 2.7. The critical length L both for 50% and 3 dB drop are marked.

Exercise 2.7

Which power drop (in dB) can we expect in 1 m normal glass (α'=4600 dB/km) and in 1 m special glass for fibers (α'=0.05 dB/km)?

▶ **Tip**
   Help H2-7 (Sect. 12.1)
   Solution S2-7 (Sect. 12.2)
   G:\Geteilte Ablagen\ST GLS Projekte\1-Frontlist_A-E\Brückner, Elements of Optical Networking\WeTransfer\Solutions.docx—S2-2

We will encounter the term engineering power again in Chapter 5 (Laser).

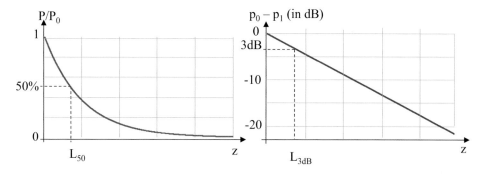

Fig. 2.7 Power drop in mW (a) and power drop in dB (b) in the z-direction

2.4 Analogue or Digital, Bandwidth or Bit Rate

All our hearing and seeing is analogue, i.e., continuously in time and amplitude. The older citizens have grown up with analogue transmission technology (telephone, television). In analogous technology, every tone, every noise, and every music is the superposition of many different sinusoidal waves with a certain frequency f_s and amplitude A_s. For example, sounds of a tuning fork (with a given frequency of 440 Hz) and of a coffee grinder are depicted in Fig. 2.8. The result is an apparent confusion in the temporal development of the amplitude A(t), usually indicated between -100% and $+100\%$.

Using the Fourier transformation, the frequency image or spectrum S(f) is obtained from the time image, usually given in decibels (dB). While the sound of the tuning fork can still be reasonably described as a sine wave (Fig. 2.8a), the coffee grinder delivers a rather chaotic temporal development of the amplitude (Fig. 2.8b). As you can see, there is no such thing as a "pure" sine wave, but every sound consists of several frequencies. Thus, each noise or sound occupies a certain frequency range—the frequency spacing of the 3 dB drop is called the spectral bandwidth Δf. Due to the frequency composition, the tone "e" of the tuning fork differs, for example, from the same tone "e" of the piano or violin.

It was not until the 1950s that digital transmission technology emerged. To do this, the analogue signal first had to be completely quantised and converted into a digital signal, that means, both the time and the amplitude must be digitized. This happens in an AD converter. Figure 2.9 sketches the mode of operation.

First, the timeline is digitalised. Sampling points in time t_s are defined on the time axis t in accordance with the so-called sampling theorem [Brü 22]: The sampling frequency or sampling rate f_s must be at least twice as large as the highest frequency to be transmitted f_{max} ($f_s \geq 2f_{max}$). The maximum spacing of the sampling points Δt_s (2.9a) is

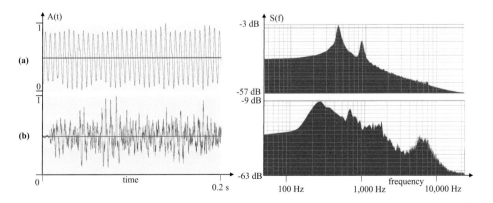

Fig. 2.8 Time course A(t) and spectrum S(f) of a tuning fork with the sound "e" or "mi" (a) and the sound of a coffee grinder (b)

then the reciprocal of the sampling rate f_s ($\Delta t_s = \frac{1}{f_s}$). For example, f_{max} for the transmission of sounds in the human range is 20 kHz, then the distance between the sampling points must be at most $\Delta t_s = \frac{1}{2 \cdot 20\ kHz} = 25\ \mu s$.

Afterwards, digitize the amplitude: To do this, the entire amplitude range is divided into individual segments, e.g., in eight segments from segment 0 to segment 7 (Fig. 2.9a). At each sampling point t_s, the segment in which the analogue signal is located is read. The segment number (e.g., Fig. 2.9a) is now represented digitally, in this case "3" as "011". So, we use 3-bit digitization in our example. This 3-bit representation of the amplitude is now placed in the range Δt_s before the sampling point t_s. This gives us the analogue signal in digital form (Fig. 2.9b).

We will encounter this procedure again in Chapter 6 (modulation) under the term pulse-code modulation (PCM).

And don't forget - at the end of the transmission, a DA converter must produce an analogue signal again, because only analogue signals can be processed in the humanitarian field.

2.5   Bits and Bytes, Bit Sequences

Usually, bits are considered as rectangular pulses (Fig. 2.10a). Since there are no discontinuities (e.g., corners) in physics, it is better for mathematical consideration to view the bit as a Fourier series or a super-Gaussian pulse (Fig. 2.10b and c). Extremely short pulses (100 ps or shorter) *must* then be regarded as Gaussian-shaped pulses (Fig. 2.10d).

The dual number system knows only two states: 0 or 1, off or on. This simplicity leads to the term *bit*. In the following, we will consistently use the small letter b as an abbreviation for bit. Caution: This is not done in the same way in all textbooks and articles!

8 Bits Form a Byte as a kind of "Word". For Byte we use the capital letter B.

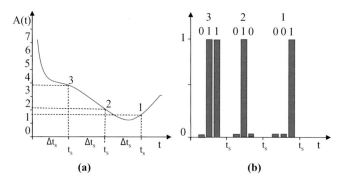

Fig. 2.9 From analogous (**a**) to digital signals (**b**)

The so-called half-value width of the bits τ_B and the distance between neighboring bits Δt_b are important. Bit rate BR is the number of bits per time unit.

$$BR = \frac{1}{\Delta t_b} \tag{2.16}$$

For the example given in Fig. 2.9, one obtains $BR = \frac{3\ Bit}{25\ \mu s} = 120$ kilobits per second (kbps). These are the quantities well known from the smartphone, such as megabits per second (Mbps) or gigabits per second (Gbps). The bit rate is thus a characteristic variable in digital transmission technology.

In digital transmission technology, every letter and every pixel are now encoded. A sequence of bits is created. There are two possibilities for the representation of the bits, the so-called formats (Fig. 2.11):

- Within the bit spacing Δt_B, the pulse always remains at the same level, one or zero. This format is called non-return-to-zero (NRZ, Fig. 2.11a).
- Only part of the bit spacing is at the one or zero level, the rest of the range remains at the zero level. This is the so-called return-to-zero format (RZ, Fig. 2.11b). In practical transmission technology, this format is mostly used.

In digital *optical* data transmission, the data are in specific bit sequences with a specific *optical power* P (Fig. 2.12).

Each single bit (see also corresponding pdf file) is like a light pulse of a certain pulse duration τ_P determined by the corresponding transmission rate. In this case, instead of the averaged power P, we have to use the *peak power* $\hat{P}$ or the *pulse energy* $E = \hat{P} \cdot \tau_P$; unit is Joule (J) or Ws. Another often-used unit is *electron volt* (eV):

$$E(\text{in eV}) = 1,6 \cdot 10^{-19} E(\text{in Ws}) \tag{2.17}$$

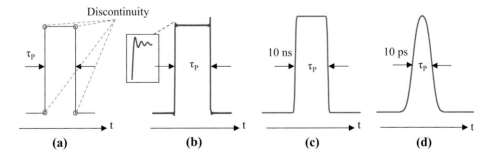

Fig. 2.10 Rectangular pulse (**a**), Fourier series pulse (**b**), Super-Gaussian (**c**), and Gaussian pulses (**d**) of duration τ_P

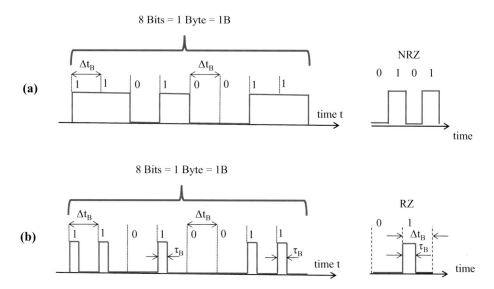

Fig. 2.11 One byte in NRZ (a) and RZ format (b)

Exercise 2.8

How much is the pulse energy of a bit with averaged power P = 100 mW and a pulse duration $\tau_P = 1$ ns?

▶ **Tip**
  Help H2-8 (Sect. 12.1)
  Solution S2-8 (Sect. 12.2)
  G:\Geteilte Ablagen\ST GLS Projekte\1-Frontlist_A-E\Brückner, Elements of Optical Networking\WeTransfer\Solutions.docx—S2-2

Light consists of photons with the *photon energy* h·f, the number of photons N inside a bit with pulse energy E can be calculated by

$$N = \frac{E}{h \cdot f} \tag{2.18}$$

where h·f is the energy of a single photon.

Fig. 2.12 Example of bit pattern

Exercise 2.9

How many photons of photon energy 1 eV (corresponding to a wavelength of about 1300 nm) are in a bit with the averaged power $P = 100$ mW and the pulse duration $\tau_P = 1$ ns?

▶   **Tip**

Help H2-9 (Sect. 12.1)
Solution S2-9 (Sect. 12.2)
G:\Geteilte Ablagen\ST GLS Projekte\1-Frontlist_A-E\Brückner, Elements of Optical Networking\WeTransfer\Solutions.docx—S2-2

For receiving a signal (bit), a receiver needs only a few photons. Thus, we get an idea of the minimum necessary bit energy required and therefore, of the limits of optical transmission technology.

Glass Fibers

3

3.1   Light Guiding in Waveguide Structures

Everyone is familiar with light propagation in free space - e.g., in the air. Message transmission with the help of light via a direct line of sight is also called optical directional radio or Free Space Optics (FSO). Systems using infrared laser diodes (wavelength 850, 1300, or 1550 nm) achieve transmission distances of up to about 4 km and data transmission rates of over 2.5 gigabits per second (Gbps) with high availability, i.e., for building-up areas. Range and availability are limited by absorption and scattering by aerosols as well as beam interruptions (e.g., by bird flight). *Optical free-space transmission* (FST) is used for the coupling of fast computer networks within or between different buildings; it is still used today as a permit-free data transmission.

In optical transmission technology, waveguiding structures are used for data transmission with light. Some examples of this are depicted in Fig. 3.1.

Guided waves or waveguiding can only be achieved if n_1 and $n_3 < n_2$. This is the case in the examples of Fig. 3.1—sandwich structure (Fig. 3.1a), buried waveguide (Fig. 3.1b), and glass fiber (Fig. 3.1c). Specifically, waveguiding in an optical fiber (Fig. 3.1c) is only achieved if the refractive index in the core n_2 is greater than in the cladding n_1. If the difference between core and cladding refractive index is small, one speaks of *weak guidance* of the light.

How do you achieve a change in the refractive index in the core since the core and cladding are made of glass? Glass fibers consist of a core with the core diameter d and the refractive index $n_2 = n_{co}$ and a cladding with the diameter D and the refractive index $n_1 = n_{cl} < n_{co}$. The change in the refractive index is achieved during the manufacture of the glass fiber by replacing a certain percentage of the glass (SiO_2) with another material.

Supplementary Information The online version contains supplementary material available at https://doi.org/10.1007/978-3-658-43242-3_3.

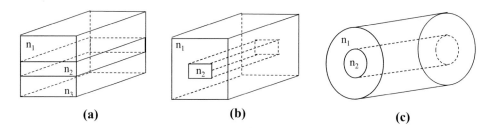

Fig. 3.1 Waveguide structures: Sandwich structure (a), buried waveguide (b), and glass fiber (c)

To increase the refractive index, e.g., GeO_2, $TiO_{,2}$ or P_2O_5 are used, to reduce it, e.g., B_2O_3 or fluorine F. The dependence of the refractive index on the concentration of the additives is shown in Fig. 3.2 using data from [Hul 96]. The reference wavelength in Fig. 3.2 is $\lambda = 1500$ nm (n $= 1.445$ in SiO_2).

Exercise 3.1

Which core material should be chosen so that the refractive index in the core exceeds the refractive index of the cladding by 0.02 (let the cladding material be SiO_2)?

▶   **Tip**
    Help H3-1 (Sect. 12.1)
    Solution S3-1 (Sect. 12.2)

3.1.1   Wave Guiding in Layers

A sandwich structure consists of a light-guiding layer with the refractive index n_2. Above and below, there is a substrate (layer III, refractive index $n_3 < n_2$) and a superstrate (layer I, refractive index $n_1 < n_2$). To consider the light propagation in a sandwich structure, we have to analyze as it is sketched in Fig. 3.3:

Fig. 3.2 Change in the refractive index of glass

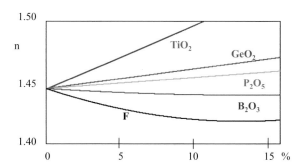

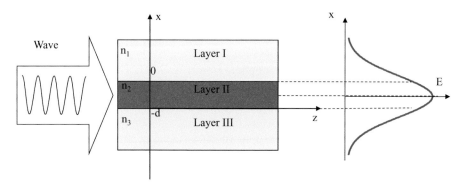

Fig. 3.3 Sandwich structure from Fig. 3.1a with cosine-shaped electromagnetic wave

One can determine the distribution of the electric field in the x-direction (see Fig. 3.3) from the wave equation (Eq. (2.6) in Chap. 2) by writing the wave equation for each layer separately (the difference is only in the respective refractive index):

$$\begin{cases} \text{in layer I} : \Delta \vec{E}\left(\vec{r},t\right) - \frac{n_1^2}{c^2}\frac{\partial^2 \vec{E}\left(\vec{r},t\right)}{\partial t^2} = 0 \\ \text{in layer II} : \Delta \vec{E}\left(\vec{r},t\right) - \frac{n_2^2}{c^2}\frac{\partial^2 \vec{E}\left(\vec{r},t\right)}{\partial t^2} = 0 \\ \text{in layer III} : \Delta \vec{E}\left(\vec{r},t\right) - \frac{n_3^2}{c^2}\frac{\partial^2 \vec{E}\left(\vec{r},t\right)}{\partial t^2} = 0 \end{cases} \tag{3.1}$$

Laplace operator Δ will be considered here in Cartesian coordinates (x, y, z), i.e.,

$$\Delta = \frac{\partial^2}{\partial x^2} + \frac{\partial^2}{\partial y^2} + \frac{\partial^2}{\partial z^2}$$

Now we look for a *joint solution* $\vec{E}$ (x) of Eq. (3.1). which is valid for *all three* layers. To this end, we have to solve equation Eq. (3.1) for each layer separately (E_I; E_{II}; E_{III}). Then we can find the parameters from the boundary condition: At x=0 and x=−d (Fig. 3.3), values of $\vec{E}$-field should be equal and continuous, i.e.,

$$E_I(x = 0) = E_{II}(x = 0) \quad \text{und} \quad \frac{E_I(x = 0)}{dx} = \frac{E_{II}(x = 0)}{dx}$$

and

$$E_{II}(x = -d) = E_{III}(x = -d) \text{ und } \frac{E_{II}(x = -d)}{dx} = \frac{E_{III}(x = -d)}{dx}$$

For example, for the symmetrical case $n_1 = n_3$, one can obtain a solution where the $\vec{E}$ field occurs in all 3 layers (see Fig. 3.1a and b). As can be seen, an electric field strength occurs not only in the waveguiding layer II but also in the substrate and the superstrate - all layers contribute to waveguiding. This case occurs in the calculation of waveguiding structures, which occur, e.g., in Chap. 4.

3.1.2   Wave Guiding in Glass Fibers

Since optical fibers are a main element of optical transmission technology, this - consideration shall now be transferred to an optical fiber (Fig. 3.4c). The glass fiber generally consists of a core material with the core diameter d (or radius a) and the refractive index n_2 and a cladding with the outer diameter D and the refractive index n_1 (Fig. 3.4a).

In contrast to layer waveguides, we are now dealing with a radially symmetrical coordinate system and thus with cylindrical coordinates (radial component r, azimuth angle φ, and propagation direction z). The Laplace operator is then:

$$\Delta = \frac{\partial^2}{\partial r^2} + \frac{1}{r}\frac{\partial}{\partial r} + \frac{1}{r^2}\frac{\partial^2}{\partial \phi^2} + \frac{\partial^2}{\partial z^2}$$

Consequently, the wave equation must be written in cylindrical coordinates r, φ, and z:

$$\left(\frac{\partial^2}{\partial r^2} + \frac{1}{r}\frac{\partial}{\partial r} + \frac{1}{r^2}\frac{\partial^2}{\partial \phi^2} + \frac{\partial^2}{\partial z^2}\right)E(r,\phi,z) + k_0^2 n^2(r)E(r,\phi,z) = 0 \qquad (3.2)$$

With the propagation direction z, one uses a separation approach for the radial component

$$E(r,\phi,z) = E_0 \cdot f(r) \cdot e^{il\phi} \cdot e^{-i\beta z} \qquad (3.3)$$

with β as the propagation constant of the wave ($\beta = k_0\,n_{eff}$), n_{eff} - effective refractive index (it acts "together" in core and cladding), and $\vec{E}_0$ - amplitude. $l = 0, 1, 2\ldots$ is the azimuth number. For guided waves, the following must be applied: $n_2 \cdot k_0 < \beta < n_1 \cdot k_0$. So, the first thing to calculate would be n_{eff}.

Using the same beginning, we get two differential equations: one for core and one for cladding regions.

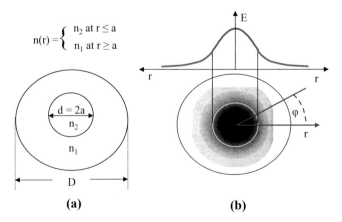

Fig. 3.4 Structure (a) and field distribution (b) in an optical fiber

$$\text{core} : \frac{\partial^2 f}{\partial r^2} + \frac{1}{r}\frac{\partial f}{\partial r} + \left(k_0^2 n_2^2(r) - \beta^2 - \frac{l^2}{r^2}\right)f = 0$$
$$\text{cladding} : \frac{\partial^2 f}{\partial r^2} + \frac{1}{r}\frac{\partial f}{\partial r} + \left(k_0^2 n_1^2(r) - \beta^2 - \frac{l^2}{r^2}\right)f = 0 \tag{3.4}$$

Using $\beta = k_0 n_{eff}$ one gets:

$$\text{core} : \frac{\partial^2 f}{\partial r^2} + \frac{1}{r}\frac{\partial f}{\partial r} + \left(k_0^2 n_2^2(r) - k_0^2 n_{eff}^2 - \frac{l^2}{r^2}\right)f = 0$$
$$\text{cladding} : \frac{\partial^2 f}{\partial r^2} + \frac{1}{r}\frac{\partial f}{\partial r} + \left(k_0^2 n_1^2(r) - k_0^2 n_{eff}^2 - \frac{l^2}{r^2}\right)f = 0 \tag{3.5}$$

With the abbreviations

$$\alpha^2 = n_2^2 - n_{eff}^2$$
$$\gamma^2 = n_{eff}^2 - n_1^2$$

one gets two differential equations for core and cladding regions, respectively:

$$core(r \le a) : \left(\frac{\partial^2}{\partial r^2} + \frac{1}{r}\frac{\partial}{\partial r} + \left(k_0^2 \alpha^2 - \frac{l^2}{r^2}\right)\right)f(r) = 0 \tag{3.6}$$

$$cladding(r \ge a) : \left(\frac{\partial^2}{\partial r^2} + \frac{1}{r}\frac{\partial}{\partial r} + \left(k_0^2 \gamma^2 - \frac{l^2}{r^2}\right)\right)f(r) = 0 \tag{3.7}$$

Both equations are so-called *Helmholtz equations*, developed by Hermann Ludwig Ferdinand von Helmholtz (1821–1894). Their solution has long been well known in mathematics: the *Bessel functions* as the solution of *Bessel's differential equations*.

As the solution of Eq. (3.6) for $r \le a$ one gets $J_l(k_0 \alpha r)$, the *Bessel function first kind* and *order* l. The differential Eq. (3.6) for $r \ge a$ can be transformed to the first type of equation using an imaginary argument. One gets $K_l(k_0 \gamma r)$, the *modified Bessel function second kind* and *order* l. Thus, $K_l(k_0 \gamma r)$ drops exponentially at large $k_0 \gamma r$.

Then the total solution is

$$f(r) = \begin{cases} c_1 \cdot J_l(k_0 \alpha r) & \text{if } r \le a \\ c_2 \cdot K_l(k_0 \gamma r) & \text{if } r \ge a \end{cases} \tag{3.8}$$

We can again get constants c_1 and c_2 from the requirement that $\vec{E}$-Fields of core and cladding should have equal values and should be continuous at $r = a$:

$$E_K(r = a) = E_M(r = a) \text{ and } \frac{dE_K(r = a)}{dr} = \frac{dE_M(r = a)}{dr} \tag{3.9}$$

As the result of these calculations, one can obtain the distribution of the electric field in radial direction as it is depicted in Fig. 3.4b. Once again one can see in b that the main part of the wave propagates in the core; nevertheless, a certain part propagates in the cladding region. It results in the term "*modes*" described in Sect. 3.2.2.

3.2 Light Guiding in Glass Fibers, Multi-Mode, and Single-Mode Fibers

3.2.1 Light Propagation in Glass Fibers, Angle of Acceptance, Numerical Aperture

The wave diagram (Sect. 3.1.2) describes the transport of light comprehensively, but it is not very descriptive. The transport of light in glass fibers [Ped 08], [LTU 87] is therefore often described graphically within the framework of the "beam model". Figure 3.5 shows the coupling of a beam with a short focal length converging lens (focal length of about 5 mm) into the face of an optical fiber. As shown in Fig. 3.5, the coupling into the glass fiber takes place at different angles—the selected beams are marked in the figure with the numbers 1–4. Beam 1 (solid line in Fig. 3.5) passes through the center of the lens and spreads out exactly in the center of the core; this central straight line is also called the optical axis. Beam 2 (dashed line in Fig. 3.5) enters the fiber at the angle α; refraction occurs from the optically thinner medium (e.g., air with the refractive index $n_a = 1$) into an optically thicker medium, e.g., glass with the refractive index $n_2 = 1.5$, according to Snellius' law of refraction:

$$\frac{\sin \alpha}{\sin \alpha'} = \frac{n_2}{n_a} \tag{3.10}$$

At the interface between core and cladding, we can have a total reflection. Repeating this total inner reflection, we find "zigzag" beam propagation as depicted in Fig. 3.5.

Beam 3 (green line in Fig. 3.5) first reaches the interface air-core under the angle φ_A and we can find a refraction corresponding to Snellius' law:

$$\frac{\sin \varphi_A}{\sin \varphi_A'} = \frac{n_2}{n_a} \tag{3.11}$$

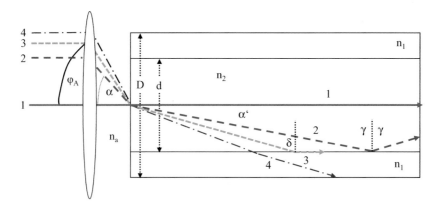

Fig. 3.5 Coupling in and beam propagation in a glass fiber

The interface core-cladding beam 3 reaches under the *critical angle of total reflection* δ, the refraction law is

$$\frac{\sin \delta}{\sin 90^o} = \frac{n_2}{n_1} \text{ with } \delta = 90^0 - \varphi_A' \tag{3.12}$$

At an angle of incidence φ_A, beam propagates theoretically exactly along the interface area core-cladding. This maximum angle of incidence φ_A, where the light is still propagating inside the core, is named the *angle of acceptance*. The sine of the angle of acceptance is the so-called *numerical aperture* NA. From the formula mentioned above, we easily find with $n_a = 1$:

$$\varphi_A = \arcsin \text{ NA} \tag{3.13}$$

$$NA = \sin \varphi_A = \sqrt{n_2^2 - n_1^2} = n_2 \sqrt{2\Delta} \tag{3.14}$$

or with the *relative* or *normalized refractive index* Δ:

$$\Delta = \frac{n_2^2 - n_1^2}{2n_2^2} \cong \frac{n_2 - n_1}{n_2} \tag{3.15}$$

The double acceptance angle is then the opening angle $\theta = 2\varphi_A$.

For glass with $n_2 = 1.5$ and $n_1 = 1.485$, we get the normalized refractive index $\Delta = 1\%$ and a numerical aperture of 0.21. Thus, the angle of acceptance is $\varphi_A = 12°$. In accordance with this beam model, that means that only beams with an angle of incidence less than the angle of acceptance $2\varphi_A$ will be guided inside the core of the glass fiber.

If the angle of incidence is more than φ_A (beam 4 in Fig. 3.5), the beam will be refracted at the interface core-cladding and propagates in the cladding. At the rough surface, the beam will be scattered and/or absorbed—the light is lost for transportation in the glass fiber.

Exercise 3.2

Calculate normalized refractive index, numerical aperture, and angle of acceptance if $n_2 = 1.48$ and $n_1 = 1.47$.

▶ **Tip**
Help H3-2 (Sect. 12.1)
Solution S3-2 (Sect. 12.2)

Exercise 3.3

The normalized refractive index is $\Delta = 0.01$, core material is SiO_2; calculate the refractive index of the core. How much is the angle of acceptance?

▶ **Tip**
Help H3-3 (Sect. 12.1)
Solution S3-3 (Sect. 12.2)

3.2.2   Transversal Modes in Glass Fibers, Mode Mixing

According to the ray model, a zig-zag propagation of light in the core can occur according to Fig. 3.5 for any angle of incidence α as long as $\alpha \le \varphi_A$. Taking into consideration the wave character of the light, one gets a different picture than in Fig. 3.5. For a detailed description of these phenomena, one has to use the wave equation. Here we will give a more descriptive picture. Taking an angle of incidence below the angle of acceptance, we have to take into consideration the interference of the wave fronts of incidence and reflected waves at transitions between core and cladding (e.g., points A and B in Fig. 3.6).

If we consider the beam with fixed phase fronts before the first reflection (dotted lines, representing phase fronts) and the beam after the second reflection (straight lines), we will get an optical retardation resulting in different positions of the phase fronts. In general, this results in a *destructive* interference and (after many reflections) in an extinction of the light. Only when the phase fronts are matched - i.e., only if the phase difference which is caused by retardation is a multiple of 2π - do we get a *constructive* interference. This model is known as the zigzag model of light propagation in a glass fiber.

As one can see in Fig. 3.6 with a core diameter d, we get the optical retardation Δl as the difference between propagation from point A to B ($\overline{AB}$) and the "virtual" forward propagation from point A to C ($\overline{AC}$):

$$\Delta l = \overline{AB} - \overline{AC} = 2d\sqrt{1 - \sin^2 \alpha}$$

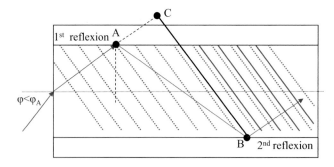

Fig. 3.6 Transversal modes in waveguides

Therefore, optical retardation results in a phase difference in accordance with Eq. (2.9):

$$\Delta\phi = k \cdot \Delta l = \frac{2\pi \, n_2 \, \Delta l}{\lambda}$$

For the total phase difference, we have to take into consideration the phase jump δ at reflection, which changes for different polarization of the light ($\perp$ or $\parallel$):

$$\tan\frac{\delta_\perp}{2} = -\frac{\sqrt{\sin^2\alpha - \left(\frac{n_1}{n_2}\right)^2}}{\cos\alpha} \text{ or } \tan\frac{\delta_\parallel}{2} = -\frac{\sqrt{\sin^2\alpha - \left(\frac{n_1}{n_2}\right)^2}}{\left(\frac{n_1}{n_2}\right)^2 \cos\alpha} \tag{3.16}$$

At vertical polarization for constructive interference, the phase difference should be

$$\Delta\varphi + 2\delta\perp = 2\pi q \text{ with } q = 0,1,2\ldots$$

therefore, a multiple of 2π:

$$2d\frac{2\pi}{\lambda}n_2 \cdot \sqrt{1 - \sin^2\alpha} - 4\arctan\frac{\sqrt{\sin^2\alpha - \left(\frac{n_2}{n_1}\right)^2}}{\cos\alpha} = 2\pi q \tag{3.17}$$

This transcendent equation can be solved only numerically. In this case, we look for equal values of the functions f_1 and f_2, where the following abbreviations are used

$$f_1 = d\frac{2\pi}{\lambda}n_2\sqrt{1 - \sin^2\alpha} \text{ and } f_2 = 2\arctan\frac{\sqrt{\sin^2\alpha - \left(\frac{n_1}{n_2}\right)^2}}{\cos\alpha} + \pi q \tag{3.18}$$

We look for values of the angle α where $f_1(\alpha) = f_2(\alpha)$:

$$f_1(\alpha) = f_2(\alpha) \tag{3.19}$$

From interface points $f_1(\alpha) = f_2(\alpha)$, we can find "permitted" angles α. This interference takes place in the x- and y-directions. The corresponding waves at these angles are named *transversal modes*. Thus, every transversal mode can be described with an exactly determined angle of incidence.

We solved Eq. (3.19) numerically using $\lambda = 1300$ nm. For simplicity, we considered only x-direction. The solution was performed for two cases:

$$d = 9 \, \mu m, \; n_2 = 1.49. \; n_1 = 1.4883;$$

therefore, $\varphi_A = 4.1°$ and $\alpha_{min} = 87.3°$ (Fig. 3.7a).

$$d = 50\mu m, \; n_2 = 1,49, \; n_1 = 1,48;$$

therefore, $\varphi_A = 9.9°$ and $\alpha_{min} = 83.3°$ (Fig. 3.7b).

At $d = 9$ μm (Fig. 3.7a), we get only one angle $\alpha_{q=0} = 89.9°$ with $f_1 = f_2$. That means only *one transversal mode* with $q = 0$ is able to propagate - this is the so-called TE_0

mode. If only a single mode is able to propagate in the fiber, the fiber is named Single-Mode Fiber (SMF). Fibers with $d=9\,\mu m$ are so-called Standard SMF (SSMF). Details on the calculation of the modes in radial direction in SMF can be found in the appendix "Extras" under "SMF.xmcd" or as pdf file.

At $d=50\,\mu m$ (Fig. 3.7b), one can find solutions for $q=0$ ... 13, i.e., 14 different transversal modes from TE_0 up to TE_{13} are able to propagate in the fiber - this is the so-called Multi-Mode Fiber (MMF). In Europe, $d=50\,\mu m$ is the standard diameter of an MMF (the US standard diameter of MMF is 62.5 μm). Thus, we found that 14 transversal modes are able to propagate in the fiber. Considering the second direction (y direction), one can expect in the fiber approximately $14^2=196$ modes. Details on the calculation of the modes in radial direction in MMF can be found in the appendix "Extras" under "MMF.xmcd" or as pdf file.

Every transversal mode corresponds to an angle α. Instead of α, one can use also the so-called *effective index* $n_{eff}=n_2\sin\alpha$, i.e., a virtual refractive index, which the light can "see" during propagation. Based on the virtual refractive index n_{eff} we can consider also the *virtual* or *effective wavelength* $\lambda_{eff}=\lambda/n_{eff}$ of the mode, which also characterizes the current transversal mode.

Results of calculations corresponding to Fig. 3.7 are summarized in Tab. 3.1. To characterize transversal modes, one can use either the angle of incidence α, or the effective refractive index n_{eff}, or the effective wavelength λ_{eff}.

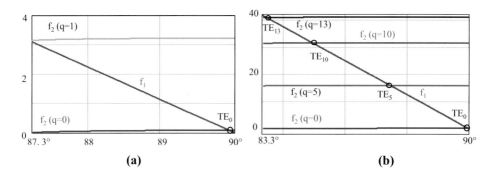

Fig. 3.7 Transversal modes in a glass fiber with $d=9\,\mu m$ (**a**) and $d=50\,\mu m$ (**b**)

Tab. 3.1 Angle of incidence, effective refractive index, and effective wavelength of SMF and MMF

SMF					MMF				
q-value	Mode	α (°)	n_{eff}	λ_{eff} (nm)	q-value	Mode	α (°)	n_{eff}	λ_{eff} (nm)
0	TE_0	89.9	1.4899	872.5	0	TE_0	88.1	1.4892	872.9
					5	TE_5	86.5	1.4878	873.8
					10	TE_{10}	84.3	1.4826	876.8
					13	TE_{13}	83.4	1.4801	878.3

Exercise 3.4

How is the number of transversal modes changed if the core diameter is reduced (give the tendency)?

▶   **Tip**
   Help H3-4 (Sect. 12.1)
   Solution S3-4 (Sect. 12.2)

The model above is based on the ray model, it describes transversal modes only qualitatively. For practical work, the cylindrical structure of an optical fiber and the associated change in the coordinate system (use of cylindrical coordinates instead of Cartesian ones) must also be considered—which, however, does not significantly change the validity of this mode picture.

By analyzing the field strength E(x,y) at the exit plane of the fiber, we can find another means of differentiation. Practically speaking, it is impossible to measure the field strength, but it is possible to measure the power distribution in the x- and y-directions by a "photo".

Thus, we have to take into consideration the relation between power P and field strength E:

$$P \propto E \cdot E^* \propto E^2 \qquad (3.20)$$

where E^* is the complex-conjugated field strength (practically it means a change in the sign at the imagined part of the field strength, e.g., instead of $-i(\omega \cdot t - k \cdot z)$ in Eq. (2.10) you have to use $+i(\omega \cdot t - k \cdot z)$).

By making "photos" of the power distribution in radial directions x and y directly at the exit plane of the fiber, one gets different distributions for different modes (Fig. 3.8). Along x- and y-directions of *transversal electromagnetically modes* (TE or TM), one gets dark points (places with $E=0$). The number of dark points n (in the x-direction) and m (in the y-direction) characterizes modes as TE_{nm}, e.g., as basic mode TE_{00} in Fig. 3.8a (no dark points in the x- and y-directions) or as TE_{10} in Fig. 3.8b (one dark point in the x-direction, no dark points in the y-direction). In real fibers, we have to use cylindrical coordinates, in which case n is the number of dark points in radial direction r and m the number of dark points at rotation around the angle $\varphi = 180°$ (see the basic mode TE_{00} in Fig. 3.8a, the next higher mode TE_{10} in Fig. 3.8b, higher modes TE_{52} in Fig. 3.8c and TE_{37} in Fig. 3.8d). In practice, however, it is usually not possible to "photographically" distinguish between these different fashions - mostly we have an interference pattern of different modes.

Up to now, we have assumed a constant refractive index of the core. These fibers are the so-called *step-index* fibers (SI).SMFs are dominantly step-index fibers. Today commercially available MMFs mostly have a parabolic index profile, i.e., we have a

Fig. 3.8 Photos and field distribution of transversal modes (photos from [Vog 02])

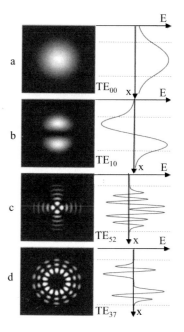

maximum index in the middle of the core and the index is reduced to the cladding in a parabolic manner. These are the so-called *gradient-index* fibers (GI).

The refractive index curve in the radial direction is generally described by the equation

$$n^2(r) = n_K^2 \left[1 - 2\left(\frac{r}{a}\right)^p \cdot \Delta \right] \tag{3.21}$$

with p as profile factor and Δ as the normalized refractive index Eq. (3.15).

As shown in Fig. 3.9 for $n_2 = 1.49$ and $n_1 = 1.48$, depending on the size of the profile factor p, different refractive index profiles are obtained from step shape (SI, $p \to \infty$) via parabolic shape ($p = 2$) to triangle shape ($p = 1$).

Figure 3.10 shows the refractive index profile of the common fiber types.

SI fibers mainly appear as SMF with $d \approx 9$ µm in optical communication, GI fibers on the other hand as MMF in Europe with $d = 50$ µm. Note, that the US standard for core diameter of GI-MMF is 62.5 µm. Details on the calculation of the modes in radial direction in step-index (SI) fibers can be found in the appendix "Extras" under "Mode field in SI fibers.pdf".

For GI fibers, $p = 2.0 \dots 1.9$ is required by ITU. In this case, we can expect more sine-shaped beam propagation (instead of the zigzag-shaped one). Because we find the highest transmission rate in SMFs, we want to look at MMFs only very briefly.

We have already mentioned that constructive or destructive interference takes place only after many reflections at the core-cladding interface. Therefore, at the beginning of

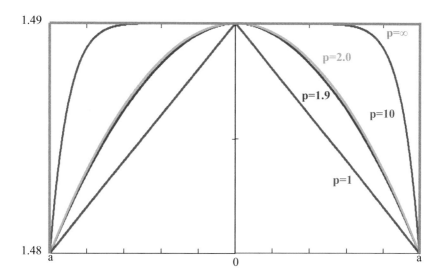

Fig. 3.9   Refractive index profiles

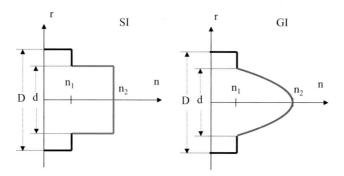

Fig. 3.10   Refractive index in SI and GI fibers

the fiber, the energy distribution over the modes is random. Only after a certain number of reflections and/or fiber bending and/or impurities in the fiber do, we find a *modal conversion* (redistribution of energy from one fashion to another): changing zero-order modes to higher order - up conversion, and vice versa - down conversion; redistribution of energy from one mode to another (Fig. 3.11).

After modal conversion, one gets a stable distribution of energy among the modes—a *stationary mode distribution*. For the mathematical treatment of this problem, however, one usually uses a so-called *equilibrium mode distribution* (EMD), i.e., that the power distribution among the modes remains stable. However, the modes have different powers. With the *unique mode distribution* (UMD), the same power is excited in all modes.

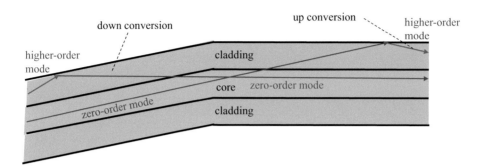

Fig. 3.11 Modal conversion

In many practical cases, especially in measurement technology, it makes sense to create a stationary mode distribution (mode equilibrium). This is done in mode mixers, in which one essentially uses fiber bends or inhomogeneities. In practice, the mode mixer used is:

- an approximately 100 m long leader fiber.
- a mandrel winding filter (approximately five glass fiber windings around a mandrel with a diameter of 12 mm).

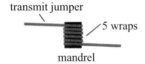

- a spherical bed mixer (pressing on spheres with 1–6 mm diameter).

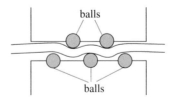

- the 70% illumination (only 70% of the core diameter or 50% of the core area and the numerical aperture are used for light transport).

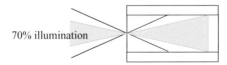

Equal distribution of optical power across all modes (mode equilibrium) is achieved by splicing together short pieces of step, gradient, and step-index fibers (SGS splices).

Modes that cannot propagate can be "soothed" with a mode scraper (in the simplest case a bare fiber consisting of core and cladding, the cladding of which is surrounded by a gel).

In engineering terms, the number of transverse modes in a glass fiber can be calculated with the help of the *fiber parameter* V. The fiber parameter characterizes the relationship between core diameter, wavelength, and NA:

$$V = \frac{\pi \cdot d}{\lambda} \cdot NA \tag{3.22}$$

With the number of the above-described example (Fig. 3.7), one gets at 1300 nm for SMF $V_{SMF} = 1.55$ and for MMF $V_{MMF} = 3.74$, respectively.

The number of transversal modes M which are able to propagate in the fiber can be estimated by

$$M_{SI} = \frac{1}{2} \cdot V^2 \qquad\qquad \textit{for SI fibers}$$
$$M_{GI} = \frac{1}{2} \cdot V^2 \left(\frac{p}{2+p}\right) = \frac{1}{2} \cdot M_{SI} \textit{ for GI fibers} \tag{3.23}$$

where p is the index profile factor (p=2 is typical for GI fibers).

For the above-described examples, one gets for SMF $M_{SMF} = 1.2$ (<u>one</u> single basic mode with orthogonal or parallel polarization), for MMF with g=2, one gets $M_{MMF} = 4$.

Exercise 3.5

Which maximum core refractive index of an SMF (cladding SiO_2) is allowed at d=5 μm and to keep the fiber parameter below 2.405?

▶ **Tip**
Help H3-5 (Sect. 12.1)
Solution S3-5 (Sect. 12.2)

3.2.3   Single-Mode-Condition, Cut-Off-Wavelength in Glass Fibers

From theoretical considerations, we can get the *single-mode-condition* as $V < V_c$, with $V_c = 2.405$ (critical fiber parameter). Thus, at fixed wavelength and numerical aperture, we can calculate the core diameter for propagation of a single mode:

$$d \le 0.766 \frac{\lambda}{NA} \tag{3.24}$$

On the other hand, at given values of d and NA, one can get the *cut-off-wavelength* λ_c:

$$\lambda_c = \frac{\lambda \cdot d \cdot NA}{2.405} \tag{3.25}$$

A single-mode operation is only possible at $\lambda > \lambda_c$.

Exercise 3.6

Is it possible to use a standard SMF (d$=9$ μm, NA$=0.1$) at $\lambda = 650$ *nm* in single-mode regime? How much is the cut-off wavelength?

▶   **Tip**
Help H3-6 (Sect. 12.1)
Solution S3-6 (Sect. 12.2)

3.2.4   Mode Field Diameter

Each mode occupies a certain space in the fiber. For example, the basic mode propagates mainly in the central part of the core (see Fig. 2.5, see also Fig. 3.4b). The distribution of the field strength (or the power corresponding to Eq. (3.20)) in radial directions (x or y) is bell-shaped (or Gaussian-shaped after the German mathematician Carl Friedrich Gauss). Therefore, the power distribution in radial direction r is

$$P(r) = E_0^2 \cdot e^{-2\frac{r^2}{w_0^2}} \tag{3.26}$$

with the mode-field diameter w_M or mode-field radius w_0:

$$w_0 = \frac{w_M}{2} \approx 2.406\frac{d}{2V} \tag{3.27}$$

The power distribution for selected fiber parameters (or as pdf file) is shown in Fig. 3.12. The mode field diameter is defined as the value at which the power has decreased to the $1/e^2 (= 0.135)$th part of the maximum power. One can also speak of a mode field area $A_{eff} = \pi \cdot w_0^2$ in which the light propagates. A_{eff} is also referred to as the effective fiber cross section. Consequently, in multimode operation (V>2.405), the light propagates mainly in the fiber core ($w_M < d$), whereas, in single-mode operation (V≤ 2.405), it propagates mainly in the core, but also in the cladding ($w_M > d$). This "dragging along" of the light power in the cladding has consequences for the dispersion (Sect. 3.4).

Fig. 3.12 Mode field diameter w_M for different fiber parameters V

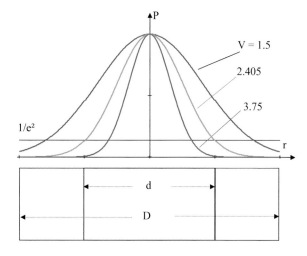

3.3   Attenuation in Glass Fibers

As mentioned in Sect. 2.3, the light power P (units W or mW) guided in the glass fiber will be reduced exponentially with the fiber length L. Power level p^{dBm} (in dBm) decreases linearly with the fiber length L^{km} (in km); thus, we get the *attenuation coefficient* α (in dB/km) as level difference with respect to 1 km fiber length:

$$\alpha = \frac{p_0^{dBm} - p_1^{dBm}}{L^{km}} \tag{3.28}$$

As we will see later, the best values for the attenuation coefficient (about 0.2 dB/km) can be found in SMF. In these fibers, we can expect—after 15 km fiber length—an attenuation of 3 dB corresponding to a ratio of 0.5 of the initial power.

Exercise 3.7

How much extra attenuation do we get in a 40 km fiber with $\alpha = 0.5\frac{dB}{km}$ compared with the best fibers?

▶ **Tip**
  Help H3-7 (Sect. 12.1)
  Solution S3-7 (Sect. 12.2)

Attenuation has an influence only on the field strength E (and therefore on the power P or the power level p, respectively). There is no influence on the light pulse duration (pulse length). There are different mechanisms for this attenuation.

3.3.1 UV Absorption

The *ultraviolet* or *UV absorption* results from a light absorption on electrons in quartz material (SiO_2) which is *not specific for wavelength*. However, the absorption is more effective the shorter the wavelength - for wavelength above 500 nm the additional attenuation by UV absorption α_{UV} is less than 1 dB/km and can be ignored (see Fig. 3.13).

3.3.2 Rayleigh Scattering

An essential process is *Rayleigh scattering* as scattering of light on microscopically small inhomogeneities in the core. Dimensions of these inhomogeneities are comparable with the wavelength of the light (maximum some µm). At these inhomogeneities, light is scattered in all directions and is mostly lost with regards to the forward direction. The increase of the attenuation coefficient α_R by Rayleigh scattering can be calculated by the formula:

$$\alpha_R = \frac{\alpha_{1\mu}}{\lambda^4} \tag{3.29}$$

where λ should be taken in µm and $\alpha_{1\mu.}$ Then the material-specific attenuation coefficient is at $\lambda = 1$ µm. For different glasses, $\alpha_{1\mu}$ is given in Tab. 3.2.

In Fig. 3.13, one can see the dependence of Rayleigh scattering on the wavelength (or as pdf file) for pure quartz glass and—for comparison—the UV absorption.

Exercise 3.8

How much extra attenuation can be expected if we use germanium-silicate glass or phosphor-silicate glass instead of silica ("poor") quartz glass a) at $\lambda = 1$ µm and b) at $\lambda = 1.5$ µm?

Fig. 3.13 Rayleigh (α_R) and ultraviolet (α_{UV}) absorption in quartz glass

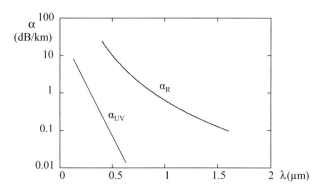

Tab. 3.2 Material constant $\alpha_{1\mu}$ for different glasses

Material	Trivial name	$\alpha_{1\mu}$ (dB/km)
100% quartz	Fused silica	0.63
Quartz + 13.5% GeO_2	Germano-silicate glass	0.8
Quartz + 9.1% P_2O_5	Phospho-silicate glass	1.1
Quartz + 25% K_2O	Valium-silicate glass	0.4
Quartz + 32.5% B_2O_3 + 16.9% Na_2O	Potassium-bore-silicate glass	1.0
Quartz + carbonate + natron	Carbonate-natron-silicate glass	1.3

▶ **Tip**
Help H3-8 (Sect. 12.1)
Solution S3-8 (Sect. 12.2)

Exercise 3.9

Typically for SMF, germanium-silicate glass is used. For special tasks (e.g., fiber amplifiers), one uses a special glass. How much extra attenuation can be expected at 1.55 μm in a 10.5 km phosphor-silicate glass?

▶ **Tip**
Help H3-9 (Sect. 12.1)
Solution S3-9 (Sect. 12.2)

3.3.3 Absorption in Water

The biggest "enemy" of glass fibers is water, especially the OH^- ions. Water molecules can be considered as dumbbells (dumbbell model) with three possible different vibrations (corresponding to different frequencies or wavelengths, respectively). These so-called basic vibrations are depicted in Fig. 3.14.

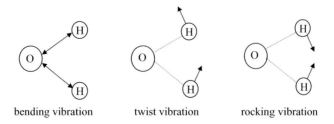

bending vibration   twist vibration   rocking vibration

Fig. 3.14 Vibrations of water molecules

Thus in water, we find specific *resonant absorption* of light (or as pdf file) at three wavelengths: 0.95 µm, 1.24 µm, and 1.395 µm (Fig. 3.14). The additional attenuation by water ions α_{OH} at a concentration of 1 ppm (water particles per million SiO_2 molecules, concentration 10^{-6}) is about 1 dB/km (at 0.95 µm) and about 40 dB/km (at 1.395 µm wavelength). This is approximately the water concentration used in Fig. 3.14. Of course, the additional attenuation depends linearly on the water concentration; at a concentration of 10^{-9}, it is 0.001 dB/km at 0.95 µm and 0.040 dB/km at 1.395 µm. In this case, additional attenuation is negligible, and we get a so-called *water-free glass fiber*.

Exercise 3.10

How much can the attenuation be reduced in an SMF at $\lambda = 1.395$ µm, if the water ion concentration is reduced from 1 ppm to 0.1 ppm?

▶   **Tip**
     Help H3-10 (Sect. 12.1)
     Solution S3-10 (Sect. 12.2)

Another type of *resonant absorption* is the absorption of light in metal ions like vanadium (V), chromium (Cr), manganese (Mn), iron (Fe), cobalt (Co), and nickel (Ni). The corresponding additional attenuation is given in Tab. 3.3.

3.3.4   IR Absorption

Glass itself consists of SiO_2 molecules. Like in water, this three-atomic molecule can perform vibrations at three wavelengths in the infrared (IR) region (9 µm, 12.5 µm, and 21 µm). The corresponding resonant absorption can be extremely strong (up to about 1010 dB/km); thus, the harmonics result in an additional attenuation down to about 1400 nm—this is the *infrared* or *IR absorption* α_{IR}.

Tab. 3.3 Extra attenuation (in dB/km) at 800 nm caused by different metal ions (concentration 1 ppm)

Metal	Substrate $Na_2O\text{-}CaO\text{-}SiO_2$	Substrate $Na_2O\text{-}B_2O_3\text{-}TlO_2\text{-}SiO_2$	Substrate SiO_2
Fe	125	5	130
Cu	600	500	22
Cr	<10	25	1300
Co	<10	10	24
Ni	260	200	27
Mn	40	11	60
V		40	2500

3.3.5   Attenuation Coefficient Versus Wavelength in Glass Fibers

Summarizing all attenuation mechanisms, one gets the attenuation coefficient versus wavelength $\alpha_{Glass}(\lambda)$ (or as pdf file). As one can see from Fig. 3.15 we get local minima, the so-called optical windows at 0.85 μm (first optical window), 1.3 μm (second optical window), and 1.55 μm (third optical window). Using optical glass fibers at these wavelengths, we get a relatively low attenuation. Nowadays, only second and third optical windows are mostly used. In the third optical window, one gets the minimum attenuation coefficient (about 0.2 dB/km). A further reduction of minimum attenuation is only possible using other fiber materials instead of glass. In today's fibers, the water content is drastically reduced; these are, for example, the so-called Low Water Peak waveguides (LWP) without water peaks (Fig. 3.15). In this case, the complete wavelength range between about 1.2 and 1.7 μm with an attenuation coefficient of less than 0.4 dB/km can be used for optical information transfer in fibers.

This bandwidth of $\Delta\lambda = 500$ nm (see Fig. 3.15) corresponds to a transmission bandwidth of about 74 THz - this is the physical limit for data transfer in glass fibers.

Today some additional bands have been added to the second and third optical windows (Tab. 3.4).

So, in summary, we find that from the point of view of absorption, the area of the third optical window is the most favorable working area with the lowest losses.

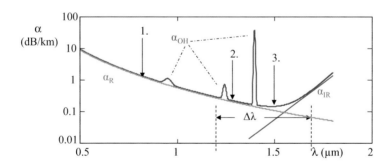

Fig. 3.15   Attenuation coefficient versus wavelength in quartz glass fibers

Tab. 3.4 Bands in optical transmission technology	Band	Name	Wavelength range (nm)
	O-Band	Original	1260–1360
	E-Band	Extended	1360–1460
	S-Band	Short	1460–1530
	C-Band	Conventional	1530–1565
	L-Band	Long	1565–1625

3.4 Dispersion in Glass Fibers

3.4.1 Concept and Impact of Dispersion Glass Fibers

The wave peak of a single frequency ω and the wave number $k = n \cdot k_0 = n \cdot \frac{\omega}{c_0} = n \cdot \frac{2\pi}{\lambda}$ propagates with the *phase velocity* $v_{Ph} = \frac{\omega}{k} = \frac{c_0}{n}$ (see Chap. 2).

If we consider the (more realistic) case of superposition of two (or many more) different monochromatic waves with different (but close to each other) frequencies ω_1 and ω_2 and wave numbers k_1 and k_2, then we get interferences or beats with an envelope instead of cosine function:

$$E(r,t) = E_0 \cdot \cos(\omega_1 t - k_1 z) + E_0 \cdot \cos(\omega_2 t - k_2 z) =$$
$$= 2E_0 \cdot \cos(\Delta\omega t - \Delta k z) \cdot \cos(\overline{\omega}t - \overline{k}z) \tag{3.30}$$

with average frequency $\overline{\omega} = \frac{1}{2}(\omega_1 + \omega_2)$ and wave number $\overline{k} = \frac{1}{2}(k_1 + k_2)$ and with the difference $\Delta\omega = \frac{1}{2}(\omega_1 - \omega_2)$ and $\Delta k = \frac{1}{2}(k_1 - k_2)$. This envelope is depicted in Fig. 3.16.

Let's consider the propagation velocity of the peak of envelope with a constant phase $(\Delta\omega \cdot t - \Delta k \cdot z = const.)$, the beat frequency (or as pdf file). Thus, we get the so-called *group velocity* v_g in a medium (here the glass fiber). v_g is determined by

$$v_g(\lambda) = \frac{dz}{dt} = \frac{d\left(\frac{\Delta\omega}{\Delta k} \cdot t - const\right)}{dt} = \frac{\Delta\omega}{\Delta k} \Rightarrow \frac{d\omega}{dk} = \frac{c_0}{n_g(\lambda)} \tag{3.31}$$

with c_0—speed of light in vacuum and n_g—*group index*. The group index n_g [Rei 97] can be calculated from the "normal" refractive index n by

$$n_g(\lambda) = n(\lambda) - \frac{dn(\lambda)}{d\lambda} \tag{3.32}$$

Therefore, the group index is given by the (wavelength-dependent) refractive index n and its changes with the wavelength.

The corresponding time delay at distance z is called the *group delay* Δt_g

$$\Delta t_g = \frac{z}{v_g} \tag{3.33}$$

Fig. 3.16 Superposition of two waves with nearly the same frequencies

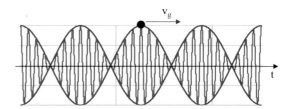

Exercise 3.11

Which index and propagation velocity is responsible for the transfer of bits?

▶ **Tip**
Help H3-11 (Sect. 12.1)
Solution S3-11 (Sect. 12.2)

When far from any resonance frequencies, the wavelength dependence of the refractive index can be approximately described by a sum of terms. It results in the so-called Sellmeier equation (see also pdf file). Mostly, the three-term Sellmeier equation is used

$$n(\lambda) = \sqrt{1 + \sum_{i=1}^{3} A_i \cdot \frac{\lambda^2}{\lambda^2 - B_i^2}} \tag{3.34}$$

where λ should be given in μm. A_i and B_i are the Sellmeier coefficients.

The first and second derivatives of the refractive index according to the wavelength are also useful—no problem for the computer. We will explain the meaning of these derivations later. First and second derivatives of the refractive index with respect to the wavelength are given by

$$\frac{dn(\lambda)}{d\lambda} = \frac{1}{n} \cdot \sum_{i=1}^{3} \frac{-A_i \lambda B_i^2}{(\lambda^2 - B_i^2)^2}$$

$$\frac{d^2 n(\lambda)}{d\lambda^2} = -\frac{1}{n^3} \cdot \sum_{i=1}^{3} \left(\frac{-A_i \lambda B_i^2}{(\lambda^2 - B_i^2)^2} \right)^2 + \frac{1}{n} \cdot \sum_{i=1}^{3} \frac{3 A_i \lambda^2 B_i^2 + A_i B_i^4}{(\lambda^2 - B_i^2)^3} \tag{3.35}$$

Coefficients A_i and B_i ($i = 1 \ldots 3$) are material-specific constants. They can be determined by measuring the wavelength dependence of the refractive index. At least six measuring points (index, wavelength) must be used to determine six unknown constants (A_i and B_i) by a least-squares fitting algorithm. In Tab. 3.5, these coefficients are given for different glass materials.

Figure 3.17 shows the dependence of the refractive index and the group refractive index on the wavelength (or as pdf file) for quartz (SiO_2) and GeO_2-doped quartz glass, respectively.

While the refractive index n decreases monotonically with increasing wavelength (the so-called "normal dispersion"), there is a minimum for the group refractive index at about 1.3 μm.

In reality, the light which appears "monochromatic" at first glance consists of many different spectral lines or wavelengths inside the line width. Each spectral line has *its own* group index and therefore *its own* group velocity. Due to different group velocities, different spectral lines need different times to pass the fiber of a certain length L - we get different running times and thus a temporal pulse broadening by dispersion.

Tab. 3.5 Specific constants for the Sellmeier formulae

material	A_1	B_2	A_2	B_2	A_3	B_3
SiO_2	0.696750	0.069066	0.408218	0.115662	0.890815	9.900559
13.5% GeO_2 +86.5% SiO_2	0.711040	0.064270	0.451885	0.129408	0.704048	9.425478
9.1% P_2O_5 +90.9% SiO_2	0.695790	0.061568	0.452497	0.119921	0.712513	8.656641
13.3% B_2O_3 +86.7% SiO_2	0.690618	0.061900	0.401996	0.123662	0.898817	9.098960
1% F +99% SiO_2	0.691116	0.068227	0.399166	0.116460	0.890423	9.993707
16.9% Na_2O +32.5%B_2O_3 +50.6% SiO_2	0.796468	0.094359	0.497614	0.093386	0.858924	5.999652

Fig. 3.17 Refractive index and group index for SiO_2 (red) and 13.5% GeO_2+86.5% SiO_2 (blue)

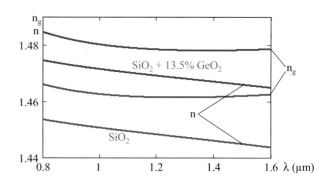

A temporal broadening of the pulse (= bit) reduces the time interval between two bits and reduces the *transfer rate* (see also pdf file). Pulse broadening corresponds to smearing of two bits at a certain fiber length L_{crit}. This situation is depicted in Fig. 3.18.

If pulse broadening t_p after a length L_1, pulses are still distinguishable. After L_2 the pulse broadening t_p is larger than the distance between two neighboring bits t_B, we cannot distinguish the bits and we are over the limit for transfer rate - information is "loosed". In this case, to achieve a high-quality transfer, we have to reduce either the transfer rate or the transfer length. Therefore, dispersion results in a pulse broadening or group delay time Δt_g which is connected with the group index n_g.

Now let us consider dispersion mechanisms in detail.

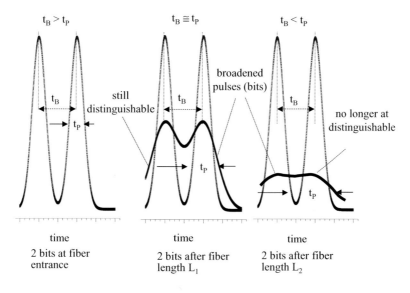

Fig. 3.18 Effect of dispersion on transmission of two bits

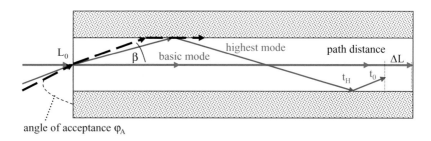

Fig. 3.19 Modal dispersion in MMF

3.4.2 Mechanisms of Dispersion

Dispersion results in a temporal pulse broadening Δt_g. There are different mechanisms; due to long distance, fiber line mechanisms in SMF are of special interest.

3.4.2.1 Modal Dispersion

As we have already discussed, the interference of light waves results in transversal modes propagating in selected directions inside the angle of acceptance φ_A. *Modal dispersion* is possible in fibers with different modes which are *able to propagate* in the fiber—that means only in MMF with a core diameter of 50 µm. The number of transversal modes decreases with decreasing core diameter; in SMF, we have only the basic modus and therefore no modal dispersion.

To determine the modal dispersion parameter $D_{Mod} = \frac{\Delta t_g}{L}$ in a step-index fiber (), we have to calculate the response time as the difference between the highest (t_H) and the fundamental mode (t_0) as $\Delta t_g = t_H - t_0$. The highest mode which is able to propagate in the fiber we can assume close to the angle of acceptance φ_A. Using Eq. (3.13), we get $\beta = arcsin\left(\sqrt{2\Delta}\right)$; the response time and the dispersion parameter are

$$\Delta t_g = \frac{n_K}{c} \cdot \frac{L_0}{\cos\beta} - \frac{n_K}{c} \cdot L_0 = \frac{n_K}{c} \cdot L_0 \cdot \left(\frac{1}{\cos\beta} - 1\right)$$

$$D^{SI}_{Moden} = \frac{\Delta t_g}{L_0} \cong \frac{n_K}{c} \cdot \Delta \tag{3.36}$$

Exercise 3.12

How much is the modal dispersion parameter of an SI-MMF if $n_2 = 1.5$ and $\Delta = 0.01$? Which maximum fiber length can be used if the run time difference should be less than 1 ns (corresponding to a bit rate less than 1 Mbps)?

▶ **Tip**
   Help H3-12 (Sect. 12.1)
   Solution S3-12 (Sect. 12.2)

For a gradient index fiber, (3.36) is changed to

$$D^{GI}_{Mod} \cong \frac{n_K}{c} \cdot \frac{\Delta^2}{2} \tag{3.37}$$

Exercise 3.13

Which modal dispersion parameter and which length do we get with numbers from Exercise 3.12 for a GI fiber?

▶ **Tip**
   Help H3-13 (Sect. 12.1)
   Solution S3-13 (Sect. 12.2)

In practice, only GI fibers are used as MMF.

3.4.2.2 Material Dispersion

Material dispersion results from the line width (spectral broadening) of the light source, here in lasers. Laser light "consists" of many different wavelengths. Due to different group indexes (see Eq. (3.32)), different wavelengths propagate in the fiber with different group velocities. If we select three different wavelengths ($\lambda_3 > \lambda_2 > \lambda_1$), at normal dispersion, we have $n_3 < n_2 < n_1$ and therefore $v_3 > v_2 > v_1$—the longer wavelength (λ_3) part of the pulse propagates the fastest in a fiber (Fig. 3.20).

Fig. 3.20 Material dispersion in glass fibers

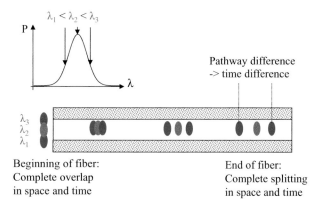

Pathway difference
-> time difference

Beginning of fiber:
Complete overlap
in space and time

End of fiber:
Complete splitting
in space and time

Dimension of material dispersion is mainly determined by *changes in the propagation time* Δt_g at *wavelength changes* $\Delta \lambda$ or by *changes of the group index* Δn_g at *wavelength changes* $\Delta \lambda$. We can calculate the material dispersion parameter D_{Mat} from changes in group speed (Eqs. (3.31) and (3.32)) with wavelength. The resulting response time at material dispersion is $\frac{t_g^{Mat}}{L} = \frac{1}{v_g}$ and changes in Δt_g^{Mat} with the wavelength is

$$\frac{\Delta t_g}{L \cdot \Delta \lambda} = \frac{d}{d\lambda}\left(\frac{1}{v_g}\right) = -\frac{\lambda}{c_0} \cdot \frac{d^2 n}{d\lambda^2} = D_{Mat} \qquad (3.38)$$

D_{Mat} is the parameter of material dispersion.

For example, using MathCad, the parameter of material dispersion D_{Mat} was calculated (Fig. 3.21) for quartz glass and for GeO_2 doped quartz glass (see also pdf file) using specific data of Tab. 3.5 and Eq. (3.35). Note that material dispersion becomes zero at $\lambda \approx 1280$ nm (SiO_2) or $\lambda \approx 1360$ nm ($SiO_2 + 13.5\%$ GeO_2), respectively.

3.4.2.3 Waveguide Dispersion
At very small core diameter, the mode field diameter is larger than the core diameter (compare with Fig. 3.12). In the cladding, the refractive index is smaller, and the speed of light is higher than in the core region. This situation is sketched in Fig. 3.22. The real

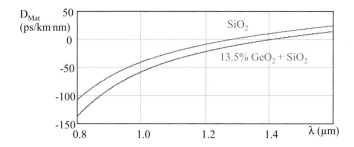

Fig. 3.21 Material dispersion in "poor" SiO_2 (red) and in 13.5% $GeO_2 + 86.5\%$ SiO_2 (blue)

Fig. 3.22 Waveguide
dispersion

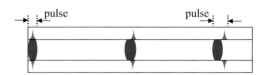

mechanism, however, is another one: The propagation speed of light pulse depends on the power distribution in core and cladding. Thus, we get a "unique" propagation speed in core *and* cladding. Because power distribution depends on wavelength, due to finite spectral width, we get again a pulse broadening.

From the above explanation, one can already see that the calculation of the waveguide dispersion D_{WG} (also as pdf file) is complicated and depends on the core diameter as well as on the core and cladding refractive indices and their structure—here only the basic idea shall be outlined.

When calculating the waveguide dispersion in an SMF, one assumes that the argument of the Bessel functions $J_0(k_0\alpha r)$ or $K_0(k_0\gamma r)$ (see Sect. 3.1.2) is constant for known parameters $k_0 = 2\pi/\lambda$, $\alpha^2 = n_2^2 - n_{eff}^2$, n_1, n_2 and $n_{eff} = \sqrt{n_2^2 - \alpha^2} = \sqrt{\gamma^2 - n_1^2}$.

Consequently:

$$\frac{2\pi}{\lambda}\sqrt{n_2^2 - n_{eff}^2} \cdot r = const. \tag{3.39}$$

From Eq. (3.39), one can calculate the dependence of the effective refractive index on the wavelength (or as pdf file) with a constant parameter of the Bessel function. With the above values ($n_2 = 1.48$, $n_1 = 1.47$, $d = 9$ μm and $n_{eff} = 1.4769$), we get

$$n_{eff} = \sqrt{n_2^2 - 0,32999 \cdot \frac{\lambda^2}{d^2}} \tag{3.40}$$

Thus, Eq. (3.40) describes the dependence of the effective refractive index on the wavelength and core diameter.

For modeling, one uses two dependencies that result from Eq. (3.40):

- The waveguide parameter $D_{WG} = \frac{\Delta t}{L \cdot \Delta \lambda}$ becomes more and more negative with increasing wavelength ($D_{WG} \sim -\lambda$): a longer wavelength "fits" worse and worse into the core area (Fig. 3.23a).
- The waveguide parameter D_{WG} increases with decreasing core diameter ($D_{WG} \sim \frac{1}{d^2}$); the mode field protrudes more and more into the cladding region (Fig. 3.23b).

One can use the following empirical formula, which reflects the above dependence:

$$D_{WG} = \frac{-320}{d^2} - \frac{3,8 \cdot (\lambda - 1,4)}{0,4} \tag{3.41}$$

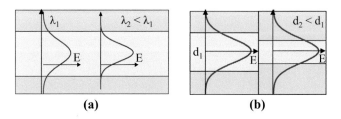

Fig. 3.23   Influence of wavelength (a) and core diameter (b) on the waveguide dispersion

As an example, Fig. 3.24 shows the material dispersion together with the waveguide dispersion at different core diameters.

Thus, with increasing core diameter D_{WG} becomes smaller, with increasing wavelength D_{WG} becomes larger (see Fig. 3.24). For this reason, waveguide dispersion only plays a role in SMF, in MMF it can be neglected. The waveguide dispersion "fortunately" has a different sign than the material dispersion (Fig. 3.24) - as a consequence, the possibility of compensating the material dispersion arises and thus the possibility of a "zero dispersion" at 1300 nm or at 1550 nm through this compensation. Details on the calculation can be found in the "Extras" folder under "WG dispersion1.pdf".

3.4.2.4 Chromatic Dispersion

Chromatic dispersion is simply the sum of material and waveguide dispersion:

$$D_{chrom} = D_{Mat} + D_{WG} \qquad (3.42)$$

As parameter of chromatic dispersion, we get

$$D_{chrom} = \frac{\Delta t_g}{L \cdot \Delta\lambda} \qquad (3.43)$$

with L as fiber length (in km) and $\Delta\lambda$ as line width (in nm); the corresponding unit is $\frac{ps}{km \cdot nm}$.

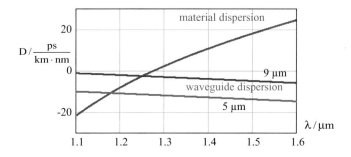

Fig. 3.24   Material dispersion and waveguide dispersion at 5 and 9 μm core diameters

The calculation proceeds in the following steps:

- Using the Sellmeier equation, calculation of core and cladding indexes $n_{co}(\lambda)$ and $n_{cl}(\lambda)$ as well as normalized index $\Delta(\lambda)$.
- Calculation of group index n_g for core and cladding $n_g(\lambda) = n(\lambda) - \lambda \cdot \frac{dn(\lambda)}{d\lambda}$.
- Calculation of material dispersion $D_{Mat}(\lambda) = -\frac{\lambda}{c_0} \cdot \frac{d^2n(\lambda)}{d\lambda^2}$ (the same for core and cladding).
- Wave equation for each area of the structure with the (common) effective refractive index $n_{eff}(\lambda)$.
- Calculation of effective index $n_{eff}(\lambda)$ from the boundary conditions for the field strength: Continuity, e.g., $E_{co}(r=r_{co}) = E_{cl}(r=r_{co})$ and differentiability, e.g., $\frac{dE_{co}(r=r_{co})}{dr} = \frac{dE_{cl}(r=r_{co})}{dr}$.
- Calculation of the field distribution in radial direction E(r), calculation of the mode field radius w_0.
- Calculation of the proportion of the wave that is transported in the core.
- Calculation of chromatic dispersion $D_{Chrom}(\lambda)$.
- Calculation of waveguide dispersion $D_{WG}(\lambda) = D_{Chrom}(\lambda) - D_{Mat}(\lambda)$.

Further details and examples for the calculation of chromatic dispersion in different glass fibers can be found in the "Extras" folder under "Chromatic dispersion".

Often in literature, the term "chromatic dispersion" of an SMF is reduced to material dispersion. In Fig. 3.25, the chromatic dispersion of pure quartz glass (as core material) with a core diameter of 9 µm (so-called standard SMF) is depicted (calculations by MathCad, or as pdf file). A zero dispersion $D_{chrom} = 0$, one can get in the second optical window at $\lambda \approx 1.31$ µm.

Summarizing we can say that material dispersion in SMF can be compensated by a suitable choice of fiber parameters (e.g., index of core and cladding, index profile, etc.) by means of waveguide dispersion. However, we have to consider that a majority of

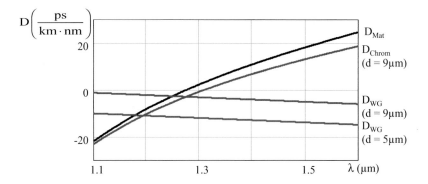

Fig. 3.25 Chromatic dispersion in poor SiO$_2$

fibers in use are standard SMF with 9 μm core diameter and zero dispersion in the second optical window.

Exercise 3.14

What is the response time Δt_g per km in a GI MMF with modal and chromatic dispersion ($n_2 = 1.5$, $\Delta = 0.01$, $D_{chrom} = 17\frac{ps}{nm \cdot km}$ and $\Delta\lambda = 1$ nm) and in an SMF with the same parameters?

▶ **Tip**
Help H3-14 (Sect. 12.1)
Solution S3-14 (Sect. 12.2)

3.4.2.5 Bandwidth-Length-Product

As one can see in Fig. 3.25, in a standard SMF, chromatic dispersion in the third optical window at $\lambda = 1550$ nm is $D_{chrom} = 18$ ps/nm·km. To distinguish two neighboring bits (see Fig. 3.18), the time distortion (i.e., mainly a pulse prolongation) should be not too big. The time distance between 2 bits t_B is connected with the bit rate BR by BR = 1/$t_B = 1/\Delta t_g$, i.e., we get

$$BR \cdot L = \frac{1}{D_{chrom} \cdot \Delta\lambda} \qquad (3.44)$$

with $\Delta\lambda$ as line width in nm.

This is the so-called Bit Rate-Length-Product (often called Bandwidth-Length-Product), which is indicated, in most cases, on the fiber by the corresponding producer. With $D_{chrom} = 18\frac{ps}{nm \cdot km}$ and a line width of $\Delta\lambda = 1$ nm, one gets

$$BR \cdot L = 55Gbps \cdot km \qquad (3.45)$$

That means that we are able to transfer 1 Gbps in a 55 km fiber or 10 Gbps in a 5.5 km SMF *without dispersion losses*.

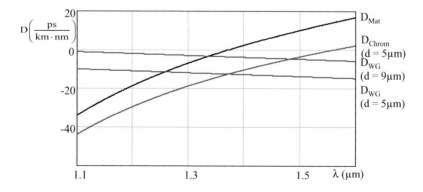

Fig. 3.26 Chromatic dispersion in GeO$_2$-doped SiO$_2$

This could lead us to consider using lasers with an extremely narrow line width (e.g., so-called DFB-MQW lasers with line width below $\Delta\lambda_L = 10^{-4}$ nm, see Sect. 5.3.2). For such narrow linewidth lasers, the modulation results in a line broadening and we get

$$\Delta\lambda = \Delta\lambda_L + BR \cdot \frac{\lambda^2}{c} \tag{3.46}$$

If we can ignore the laser line width $\Delta\lambda_L$, we get as transfer length L:

$$L \cong \frac{c_0}{BR^2 \cdot \lambda^2 \cdot D_{Chrom}} \tag{3.47}$$

with c_0 as speed of light.

Therefore, for high bit rates, it would be useful to develop fibers with decreased chromatic dispersion or to compensate dispersion.

Exercise 3.15

A SMF with $D_{chrom} = 18\frac{ps}{nm \cdot km}$ operates at $\lambda = 1.55$ μm. Over which length can one transfer a bit rate of a) 2.5 Gbps, b) 10 Gbps, or c) 40 Gbps without dispersion compensation?

▶   **Tip**
Help H3-15 (Sect. 12.1)
Solution S3-15 (Sect. 12.2)

3.4.2.6 Polarization-Mode-Dispersion (PMD)

If chromatic dispersion becomes zero, i.e., if all other types of dispersion are "switched-off" or have been compensated, in SMF one only has to consider the influence of different polarization, the Polarization-Mode-Dispersion (PMD). The corresponding mechanism is sketched in Fig. 3.27. Polarization of light in a fiber can be changed by inhomogeneities, by different conditions (e.g., stress due to small differences in core radii along the fiber), and by bending of the fiber. Even initially (at the entrance), linearly polarized light will be transferred to light polarized in both directions—and because the refractive index is slightly different for both polarizations; again, one gets a time distortion.

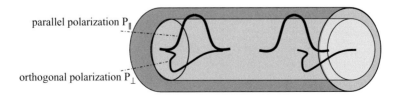

parallel polarization $P_{\parallel}$

orthogonal polarization $P_{\perp}$

Fig. 3.27 Polarization-Mode-Dispersion

Depending on the bit rate, typical PMD effects can take place in fibers of only 100 m up to over 1000 km in fiber length.

To consider the influence of PMD quantitatively, one typically assumes that response time Δt_g^{PMD} caused by PMD is not longer than 10% of the time distance between neighboring bits t_B:

$$D \cdot t_g \leq \frac{t_B}{10} = \frac{1}{10 \cdot BR}$$

The corresponding value is a statistical one; it increases with the square root of the fiber length L and with the PMD coefficient Δt_{PMD}:

$$\Delta t_g = \sqrt{L} \cdot \Delta t_{PMD} \leq \frac{1}{10 \cdot BR}$$

Thus, the transmission length L is given by

$$L \leq \frac{1}{100 \cdot BR^2 \cdot \Delta t_{PMD}^2} \tag{3.48}$$

Of course, the PMD coefficient in "older" glass fibers (installed in the 1990s) is mostly rather poor—the "very bad" fibers have $\Delta t_{PMD} \geq 2 \frac{ps}{\sqrt{km}}$, about 20% of them have $\Delta t_{PMD} = 0.8 \frac{ps}{\sqrt{km}}$. As the standard value in future, we can expect $\Delta t_{PMD} = 0.5 \frac{ps}{\sqrt{km}}$; some producers already offer now $\Delta t_{PMD} = 0.1 \frac{ps}{\sqrt{km}}$. With $0.1 \frac{ps}{\sqrt{km}}$, one can transfer 2.5 Gbps over 160,000 km, 10 Gbps over 10,000 km, or 40 Gbps over 625 km without polarization mode dispersion.

Exercise 3.16

Which transmission length can we expect in "bad" fibers ($\Delta t_{PMD} \geq 2 \frac{ps}{\sqrt{km}}$) and which in standard SMF ($\Delta t_{PMD} = 0,5 \frac{ps}{\sqrt{km}}$) for 10 Gbps?

▶ **Tip**
Help H3-16 (Sect. 12.1)
Solution S3-16 (Sect. 12.2)

The simplest way to avoid PMD seems to be the use of *polarization keeping* fibers with an elliptical core and a spherical cladding. In this case, different pressure from the cladding part (different thicknesses) favors one polarization direction (Fig. 3.28). To produce these fibers is relatively expensive, and the use of fibers which are already rolled out is impossible.

For a certain bit rate, PMD limits the transfer length for a given bit rate. After this distance, one needs a regeneration to reduce or compensate dispersion and to amplify the signals.

Fig. 3.28 Polarization
keeping fiber

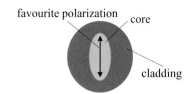

3.5   Special Glass Fibers and Fiber Cables

Depending on the objective and the task, it is possible to modify both the structure of naked fibers (core and cladding) and the entire structure of a fiber optic cable.

3.5.1   Special Fibers

It is possible to change the dispersion by varying the refractive indices or by using a special refractive index profile.

For the following special fibers, MathCad programs have been written by the author, which are listed in the "Extras" folder and lead to the corresponding links. This change in the refractive index profile compared to the standard SMF (Fig. 3.29a) occurs, for example, in dispersion flatted fiber (DFF, see Fig. 3.29b and c) or dispersion shifted fiber (DSF, see Fig. 3.29d and e). A dispersion shift analogous to DSF can also be achieved in SMFs with significantly reduced core diameters. Often, the change in dispersion is associated with a reduction in the effective mode field area $A_{eff} = \pi \cdot w_0$. To counteract this reduction, the special fiber with increased mode field area (large effective area fiber, LEAF) was developed. Another way is dispersion compensation. Through a special progression of the refractive index in the core (see Fig. 3.29f and g), it is now possible to produce glass fibers with high negative dispersion at 1550 nm that are able to compensate for chromatic dispersion (dispersion compensating fiber, DCF). The size of the dispersion in a DCF as well as the increase in dispersion (dispersion slope) can be changed by changing the ratio of the width to the core diameter. Figure 3.29 also shows the dispersion slopes of the glass fibers; they are the results of the author's calculations. Details and explanations of the calculations can be found in the "Extras" folder under "Chromatic Dispersion".

The following figure shows the course of the chromatic dispersion of these special fibers over the wavelength.

There are now many commercial suppliers of standard and special glass fibers such as Corning Glass or m2optics.

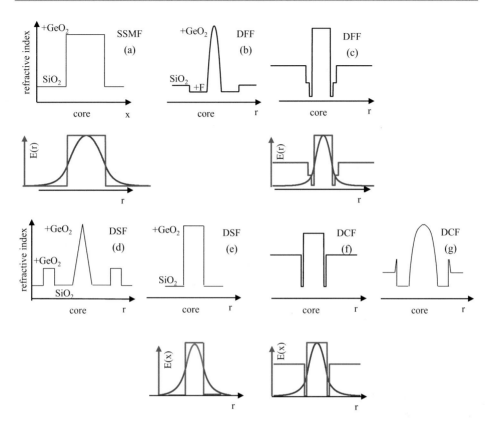

Fig. 3.29 Special fibers SSMF (a), DFF (b and c), DSF (d and e), and DCF (f and g) to change dispersion: refractive index profile n(r) and field distribution E(r)

3.5.2 Types of Glass Fibers

For data transmission over long distances, attenuation and dispersion are of particular importance. Depending on dispersion, a distinction is currently made between four single-mode glass fiber types with different refractive index profiles (see Fig. 3.29); all have a core diameter of 5–9 μm and a (standardized) cladding diameter of 125 μm; together with the coating, the outer diameter is 250 μm.

From the author's point of view, the following types of glass fibers can be distinguished. The manufacturers are generally bound to the recommendations of the International Telecommunication Union (Telecommunication Division), the corresponding

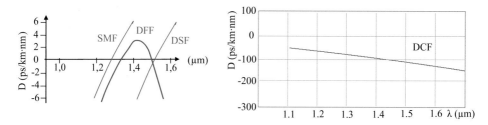

Fig. 3.30 Course of chromatic dispersion in standard SMF and in special glass fibers DFF, DSF, and DCF

recommendation numbers are given. Typical variants of the refractive index (refractive index) profile are drawn as an insert:

NDSF (**N**on-**D**ispersion-**S**hifted **F**iber); this is the standard SMF (SSMF); millions of kilometers are rolled out; there is no dispersion shift, zero dispersion at about 1300 nm (ITU recommendation G.652). As a new version, one can find the low water peak waveguides, where the typical OH⁻ peak at 1320 nm is suspended (ITU recommendation .652.C).

DSF (Dispersion-Shifted Fiber); we find a dispersion shift of zero dispersion forward to about 1550 nm, extra losses are about 0.25–0.30 dB/km (ITU recommendation G.653). Because in DSF, one can find nonlinearities (see Chap. 8), DSF can only be used for dense wavelength division multiplexing (DWDM) in limited cases.

NZ-DSF (Non-Zero Dispersion-Shifted Fiber); this fiber has a zero dispersion at margin of the third optical window. Thus, dispersion in the third optical window in accordance with ITU-T recommendation G.655 is in the range 0,1–10 ps/nm· km (True Wave RS, TeraLight). Extra losses are about 0.2 dB/km. In LEAF-fibers, the mode field is distributed over several parts of the core, resulting in a bigger area.

DCF (Dispersion Compensating Fiber); it has a strong negative dispersion in the third optical window. Extra losses are about 0.5 dB/km.

For practical use, we will summarize the most important parameters of glass fibers.

Standard Single-Mode Fiber (SSMF) and AllWave™ Fiber:

Group index of the core	1.4681
Dispersion at $\lambda = 1550$ nm	17 ps/km· nm
Dispersion slope	0.057 ps/km· nm²
Losses	0.15–0.20 dB/km
Core diameter	9 µm
Mode field diameter	9.5–11.5 µm
Effective core area	80 µm²

DSF (Dispersion-Shifted Fiber):

Dispersion at $\lambda = 1550$ nm	0 ps/ km· nm
Dispersion slope	0.07 ps/ km· nm²
Losses	0.25–0.30 dB/km

NZ-DSF (Non-zero Dispersion-Shifted Fiber) TrueWave®-RS:

Group index of the core	1.470
Dispersion bei $\lambda = 1550$ nm	4.4 ps/km· nm
Dispersion slope	0.045 ps/ km· nm²
Core diameter	6 µm
Mode field diameter	7.8–9.0 µm
Effective core area	55 µm

NZ-DSF (Non-zero Dispersion-Shifted Fiber) LEAF®:

Group index of the core	1.469
Dispersion at $\lambda = 1550$ nm	4 ps/ km· nm
Dispersion slope	0.095 ps/ km· nm²
Core diameter	6.4 µm

| Mode field diameter | 9.2–10.0 μm |
| Effective core area | 72 μm² |

NZ-DSF (Non-zero Dispersion-Shifted Fiber) TeraLight®:

Group index of the core	1.470
Dispersion at $\lambda = 1550$ nm	8 ps/ km· nm
Dispersion slope	0.058 ps/ km· nm²
Core diameter	7 μm
Mode field diameter	8.7–9.7 μm
Effective core area	65 μm²

DCF (Dispersion Compensating Fiber):

Dispersion at $\lambda = 1550$ nm	−120 ps/ km· nm
Dispersion slope	−0.2 ps/ km· nm²
Losses	0.5 dB/km
Effective core area	20 μm²

3.5.3   Glass Fiber Cables

Optical glass fiber consists of a core and a cladding region (see types of glass fibers, Sect. 3.5.2), material is in both cases doped glass. For example, standard GI-MMF has a diameter of 50 μm (EU) or 62.5 μm (US), corresponding to about 1/40 inch. The standard SMF has a core diameter of about 9 μm. The core and cladding area (the so-called naked fiber) together have a standardized diameter of 125 μm both in the EU and the USA (corresponding to about 1/20 inch). In practical fibers, the cladding is usually coated with a layer of acrylate polymer or polyimide. This coating protects the fiber from damage but does not contribute to its optical waveguide properties. At present, the diameter of the coated glass fiber is 250 μm (corresponding to about 1/10 inch, see Fig. 3.31a), but in the future, it will be 200 μm.

Buffer in optical patch fibers (Fig. 3.31a) is used to stabilize cable. Materials used to create buffers can include fluoropolymers such as polyvinylidene fluoride (trivial name Kynar), polytetrafluoroethylene (trivial name Teflon), or polyurethane.

The jacket (Fig. 3.31a) is used for sheathing and support of the fiber from mechanical and physical handling. The jacket material is application-specific. The material determines the mechanical robustness, chemical and UV radiation resistance, and so on. Some common jacket materials are low smoke zero halogen (LSZH), polyvinyl chloride (PVC), polyethylene, polyurethane, polybutylene terephthalate, and polyamide. The jacket on patch cords is often color-coded to indicate the type of fiber used (see Tab. 3.6). Tab. 3.6 also lists MMFs that are not used for communication purposes.

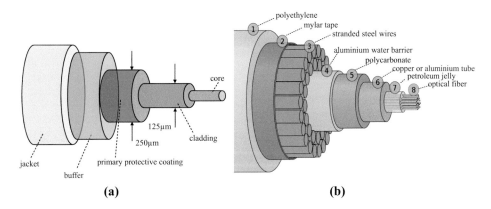

Fig. 3.31   Construction of a patch cable (a) and an optical undersea cable (b)

Tab. 3.6   Color code of different fiber types

Fiber type and class	Diameter (µm)	Jacket color
Multimode Ia (EU)	50/125	Orange
Multimode Ia (US)	62.5/125	Slate
Multimode Ia (not in communication)	85/125	Blue
Multimode Ia (not in communication)	100/140	Green
Singlemode IVa (EU)	All	Yellow
Singlemode IVb (US)	All	Red

There is a large number of different fiber optic cables, each of which considers the requirements or the specific conditions for laying. For example, the structure of an optical undersea cable (Fig. 3.31b) is strictly dependent on the construction method and the application type that determines the design and the grade of external protection. In addition to the active fibers used in optical communication, there are also unused fibers (so-called dark fibers), which are kept in reserve or, for example, used for measurement purposes. A bundle of glass fibers is surrounded by gelatinous petroleum jelly, and everything together is in a tube made of copper or aluminum (Fig. 3.31b). This is followed by a polycarbonate layer and an aluminum water barrier. Stranded steel wires provide mechanical stability. Finally, the whole is wrapped in a mylar tape and polyethylene (Fig. 3.31b).

In Fig. 3.32 an example of an undersea cable is given. Often, the strength wires are replaced by a centrally located copper wire that transports the current necessary for the amplifiers. In addition to the protective layers already mentioned, the cable may be

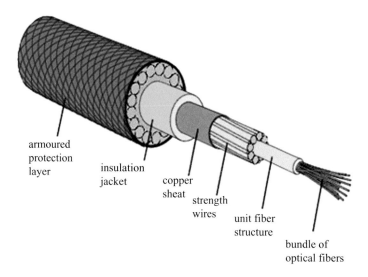

Fig. 3.32   Example of a glass fiber cable

armored to protect it from environmental hazards, such as construction work or gnawing animals. Undersea cables are more heavily armored in their near-shore portions to protect them from boat anchors, fishing gear, and even sharks, which may be attracted to the electrical power that is carried to power amplifiers or repeaters in the cable.

Modern cables come in a wide variety of sheathings and armor, designed for applications such as direct burial in trenches, dual use as power lines, installation in conduit, lashing to aerial telephone poles, submarine installation, and insertion in paved streets.

The arrangement of many SMFs is exemplified in Fig. 3.33. In Fig. 3.33a, each SMF (coated fiber) is placed into a single tube. Coated fibers can be fixed (tight fibers) or

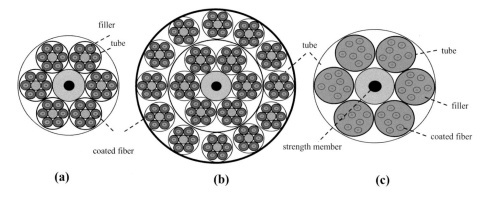

Fig. 3.33   Examples of optical fiber cables: (a) One coated fiber in a single tube, (b) multiple tube system, and (c) many loose coated fibers in a tube

loose (loose fibers) in the tube. All together we get $6 \times 6 = 36$ single-mode fibers in this cable. Instead of stranded steel wires, a strength member in the center is used.

In the multitube system of Fig. 3.33b, one gets 6×6 (inner ring) $+ 12 \times 6$ (outer ring) $= 108$ single-mode fibers.

In the cable according to Fig. 3.33c, several coated optical fibers (up to 20) are loosely housed in a tube (loose coated fibers). Thus, we have in Fig. 3.33c up to 120 glass fibers in a cable.

There are many fiber optic cable manufacturers with different applications in mind. One of the best-known companies is Corning Glass, which incidentally also developed the first low-loss glass for optical fibers.

Fiber Optic Connections and Couplers

4

An essential part of an optical network are the connectors and switches which are able to direct data fast and low loss from point A to point B, or to realize a conference involving several participants. To this end, one needs splices, plugs, couplers, and switches as well as multiplexers and demultiplexers. Its operation is described below.

4.1 Plugs and Splices

The laying of glass fibers over a long distance requires detachable connections (plugs) or non-detachable connections (splices). Because there are so many technical possibilities for plugs and splices [Hub 92, Ebe 10], we would like to focus here primarily on general aspects to consider. We have to keep in mind that any splice or plug means extra losses—for a high-quality splice of SMF, this is 0.01–0.15 dB, and for a plug, we must consider losses of 0.1–0.5 dB.

4.1.1 Splices

For splicing, well-prepared ends of fibers are fused, generally using an electrical arc discharge [Mah 94].

The following detailed steps must be performed:

- Remove the outside cladding and coating; then we get the so-called "naked fiber" which consists of core and cladding only.
- Clean the glass fiber (with ethanol or isopropanol).

© Springer Fachmedien Wiesbaden GmbH, part of Springer Nature 2024
V. Brückner, *Elements of Optical Networking*,
https://doi.org/10.1007/978-3-658-43242-3_4

Fig. 4.1 Splices

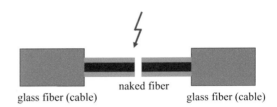

glass fiber (cable)   naked fiber   glass fiber (cable)

- Cut the glass fiber with a special tool, the so-called Miller tool. With this Miller tool, first, the fiber is scratched by a diamond grinding wheel and then it is broken. Thus, we can expect a flat break.
- To splice, we have to position the two fibers in the splicing apparatus exactly vis-à-vis in space (Fig. 4.1).
- Using a low electrical arc discharge, the naked fibers are again cleaned to remove any remaining dirt.
- Next, the front surfaces will be brought together and fused by a strong electrical arc discharge (Fig. 4.1).
- Finally, the location of fusion is covered by a shrinking hose or a metallic crimping hull to protect against mechanical damage or external influences.

4.1.2   Plugs

Plugs are *detachable* connectors for glass fibers. Therefore, when disconnecting and reconnecting plugs, there is always the danger of dirt getting in between them. To prepare a plug, we begin with the first three steps described above for splices (Sect. 4.1.1). The next important steps are grinding and polishing of the plug—only with a high-quality polishing can a physical contact take place (Fig. 4.2a). In former times to avoid Fresnel's reflection, an immersion liquid (IL) with a refractive index close to the index of glass was pressed between front surfaces (Fig. 4.2b). Both front surfaces then had to be precisely centered, e.g., an exactly matched hull with a small ventilation hole (VH).

When connecting multi-mode or single-mode fibers, there is always extra attenuation. Its value changes, depending on variations of production parameters (intrinsic losses) or quality of assembling (extrinsic losses). Additional losses are possible for uneven front surfaces (faces)—it is impossible to describe these losses mathematically. Extra losses in the coupling of multi-mode fibers are given in the following.

Fig. 4.2 Basic assembly of a plug

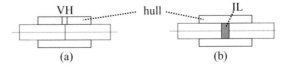

Intrinsic Losses

Critical parameter	Schema	Extra losses (dB)
Core diameter d	d_1 d_2	$a_d = -20\lg\left(\frac{d_2}{d_1}\right)$
Numerical aperture NA (angle of acceptance)	NA_1 NA_2	$a_{NA} = -20\lg\left(\frac{NA_2}{NA_1}\right)$
Profile of refractive index g SI fiber: $g=\infty$ GI fiber: $g=1{,}8 \ldots 2{,}0$	g_1 g_2	$a_g = -10\lg\left[\frac{g_2}{g_1}\left(\frac{g_1+2}{g_2+2}\right)\right]$

Exercise 4.1

How many losses can we expect when coupling an MMF according to US standards (62.5 µm) with an MMF according to EU standards (50 µm)?

▶ **Tip**
   Help H4-1 (Sect. 12.1)
   Solution S4-1 (Sect. 12.2)

Exercise 4.2

How many losses can we expect when coupling an MMF according to EU standards (62.5 µm) with an SMF (9 µm)?

▶ **Tip**
   Help H4-2 (Sect. 12.1)
   Solution S4-2 (Sect. 12.2)

Exercise 4.3

How many losses can we expect when coupling an MMF according to US standards (62.5 µm) with $g=2.0$ with an MMF according to EU standards (50 µm) with $g=1.9$? How many losses can we expect when coupling in the opposite direction (SMF -> MMF)?

▶ **Tip**
   Help H4-3 (Sect. 12.1)
   Solution S4-3 (Sect. 12.2)

Extrinsic Losses

Critical parameter	Schema	Formulae
Radial offset	o	$a_o = -10\lg\left(1 - \frac{16v}{3\pi d}\right)$
Angle error	α	$a_\alpha = -10\lg\left(1 - \frac{16\sin(\alpha/2)}{3\pi NA}\right)$
Gap	L	$a_L = -10\lg\left(1 - \frac{L}{d}NA\right)$

Furthermore, we have to consider that, at gaps $L \leq \lambda$, we can expect interference appearances. In this case, extra losses can be both smaller and bigger than had been calculated. We can check for appearances of interference by disconnecting and re-connecting the plugs several times and measuring the losses—with interference, we will get a different value each time.

Exercise 4.4

How many losses can we expect when coupling two MMFs (50 µm) with a radial offset o$=$1 µm?

▶   **Tip**
    Help H4-4 (Sect. 12.1)
    Solution S4-4 (Sect. 12.2)

Exercise 4.5

How many losses can we expect when coupling two MMFs (50 µm) with an angle error $\alpha = 2^O$ with NA$=$0.2?

▶   **Tip**
    Help H4-5 (Sect. 12.1)
    Solution S4-5 (Sect. 12.2)

Exercise 4.6

How many losses can we expect when coupling two MMFs (50 µm, NA$=$0.2) with a gap L$=$3 µm?

▶ **Tip**

Help H4-6 (Sect. 12.1)

Solution S4-6 (Sect. 12.2)

Exercise 4.7

How many losses can we expect when coupling two MMFs (50 µm) adding losses from Exercise 4.4 to Exercise 4.6?

▶ **Tip**

Help H4-7 (Sect. 12.1)

Solution S4-7 (Sect. 12.2)

If we couple SMF the mode field diameter 2w, rather than the core diameter, is of importance.

Critical parameter	Schema	Formel
Intrinsic losses		
Mode field diameter		$a_w = -20 \lg \left(\frac{2 w_1 w_2}{w_1^2 + w_2^2} \right)$
Extrinsic losses		
Radial offset o		$a_o = 4.43 \left(\frac{v}{w} \right)^2$
Angle error α		$a_\alpha = 4.34 \left(\frac{\alpha \pi n_x w}{\lambda} \right)^2$
Gap L (n_a—refractive index between fibers, with air $n_a = 1$)		$a_L = -10 \lg \dfrac{1}{1 + \left(\frac{\lambda L}{2 \pi n_s w^2} \right)^2}$

Let us calculate losses to be expected when coupling different SMFs described in Sect. 3.4.3. We assume the maximum mode field diameter in each type of fiber.

Exercise 4.8

How many losses can we expect when coupling a standard SMF ($2w_1 = 11.5$ µm) with a TrueWave® ($2w_2 = 7.8$ µm)?

▶ **Tip**

Help H4-8 (Sect. 12.1)

Solution S4-8 (Sect. 12.2)

Exercise 4.9

How many losses can we expect when coupling a TrueWave® ($2w_1 = 9.0$ µm) with a standard SMF ($2w_2 = 9.5$ µm)?

▶ **Tip**
Help H4-9 (Sect. 12.1)
Solution S4-9 (Sect. 12.2)

Exercise 4.10

How many losses can we expect when coupling a standard SMF ($2w_1 = 11.5$ µm) with a LEAF ($2w_2 = 9.2$ µm)?

▶ **Tip**
Help H4-10 (Sect. 12.1)
Solution S4-10 (Sect. 12.2)

Exercise 4.11

How many losses can we expect when coupling a LEAF ($2w_1 = 10.0$ µm) with a standard SMF ($2w_2 = 9.5$ µm)?

▶ **Tip**
Help H4-11 (Sect. 12.1)
Solution S4-11 (Sect. 12.2)

If one combines *different* SMFs (e.g., from different producers) by plugs or splices we have to consider that every transition (e.g., by splice) between two fibers with different index profiles results in different mode field diameters. Thus, one gets changes in dispersion as well as extra attenuation. The extra attenuations resulting from coupling between different fibers (see Sect. 3.5.2) are shown in Tab. 4.1. These losses are independent of the direction of connection.

Tab. 4.1 Extra losses (in dB) from connection of different SMFs

	SMF	TeraLight	LEAF	TrueWave®
SMF	0.16	0.33	0.21	0.64
TeraLight	0.33	0.05	0.08	0.20
LEAF	0.21	0.08	0.03	0.27
TrueWave®	0.64	0.20	0.27	0.09

4.1.3 Examples of Optical Fiber Connectors

In contrast to splicing, the connection of optical fibers with connectors allows fast connection and decoupling. However, this advantage is associated with some disadvantages: Connectors have higher losses (about 0.5–1 dB), the demands on mechanical accuracy are higher and due to the mechanical stress, there is a finite number of mating operations (500–1,000 cycles). The basic idea is the coupling and exact adjustment of the fiber cores. Good connectors (e.g., physical contact connectors) have very low losses due to reflection and inaccurate adjustment.

A variety of optical fiber connectors are available. In all, more than 100 different types of fiber optic connectors are on the market. The main differences among types of connectors are dimensions and methods of mechanical coupling.

Basically, a distinction can be made between four connector types:

- **SC Fiber Optic Connector:** SC stands for Square Connector or Subscriber Connector. It was developed by Nippon Telegraph and Telephone (NTT) company. SC is a snap (push-pull coupling) connector with a 2.5 mm ferrule diameter. It is standardized by the standard IEC 61,754-4. The connector's outer square profile together with its snap-coupling mechanism that allows greater connector packaging density in instruments and patch panels. With the (still) increasing popularity in the market, the manufacturing cost for SC went down. Now, it has been widely applied in single-mode fiber optic cable, analogue CAble TV or Community Antenna TeleVision (CATV), Gigabit Passive Optical Network (GPON), and Gigabit Interface Converter (GBIC).

SC connector

- **LC Fiber Optic Connector:** LC stands for Lucent Connector or Local Connector. It is a push-pull coupling, small form factor connector that uses a 1.25 mm ferrule, half the size of the SC with outer square profile. LC, due to the combination of small size and latch feature, is ideal for high-density connections, Small Form-factor Pluggable (SFP), and SFP plus transceivers as well as for 10 Gbit small form-factor pluggable (XFP) transceivers. Along with the development of LC-compatible transceivers and active networking components, it will continue to grow in the Fiber-To-The-Home (FTTH) arena.

LC connector

- **FC Fiber Optic Connector:** FC is short for Ferrule Connector. It is a round, threaded fiber optic connector that was designed by Nippon Telephone and Telegraph in Japan. The FC connector is applied for single-mode optic fiber and polarization-maintaining optic fiber. The FC is a screw-type connector with a 2.5 mm ferrule, which was the first fiber optic connector to use a ceramic ferrule. However, FC is becoming less common and mostly replaced by SC and LC because of its vibration loosening and insertion loss.

FC connector

- **ST Fiber Optic Connector:** ST stands for Straight Tip or Bayonet Fiber Optic Connector (BFOC). The ST connector was developed by AT&T shortly after the arrival of the FC. They may be mistaken for one another, but ST uses a bayonet mount other than a screw thread. And you have to make sure SC connectors are seated properly owing to their spring-loaded structure. SC is mainly used in multimode fiber optic cables, campuses, and buildings.

ST connector

SC-SC connector

With a simplex coupling, optical fiber male connectors can be easily connected to each other, e.g., by SC-SC or ST-ST connector. The coupling enables a direct connection to the optical fiber cable and is suitable for simple mounting in optical fiber connection boxes, cabinets, and patch panels.

ST-ST connector

The introduction of fiber patch cables has realized the data communicating at a much higher data rates. Of course, for the individual connector types, there are patch cables that convert to the respective other type—i.e., LC to LC fiber cable, single-mode LC to ST fiber patch cable, single-mode fiber ST to SC patch cord, single-mode fiber SC to FC connector, single-mode fiber cable with LC connector, etc.

At first glance, an optical connector hardly differs from a high-frequency electrical one (e.g., from a screwable BNC connector)—but in the case of the optical connector, access to the optical fiber is protected by a protective cap. To make a connector, prepare the optical cable in the same way as for splicing (see Sect. 4.1.1).

SC and LC connectors are the most common types of connectors on the market. Connectors exist for single cables, as well as for two (e.g., SC-DC or Dual Contact), four (e.g., SC-QC or Quattro Contact), or more (MPO or Multiple-fiber Push-On/Pull-Off) cables in one connector.

Generally, organizations like ITU will standardize on one kind of connector, depending on what equipment they commonly use.

4.2   Functionality of Couplers and Switches

4.2.1   Coupling Elements

In general, the coupler combines light streams of the same wavelength from different channels (coupler in the narrower sense) or splits them into different channels (splitter) [Hub 92, Ebe 10]. If the luminous flux at ports 1 and 2 has different wavelengths (Fig. 4.3a), which are mixed at port 3, one has a multiplexer (MUX) of the wavelengths, also called wavelength division multiplexing (WDM). If the light stream at port 1 consists of different wavelengths (Fig. 4.3b) and these wavelengths are separated to port 3 or port 4, we have demultiplexing (DEMUX).

Couplers are typically described in terms of the so-called theory of quadruples (two ports at Left Hand Side, two ports at Right Hand Side, see Fig. 4.3). As a result, we get some parameters for couplers (corresponding to Fig. 4.3a) and splitters (corresponding to Fig. 4.3b).

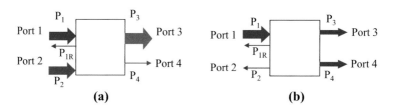

Fig. 4.3 Function of couplers (a) and splitters (b)

Coupler	Splitter

insertion losses a_i: total power at ports $3+4$ is less than at ports $1+2$:

$$a_i = -10 \cdot \log \left(\frac{P_3}{P_1 + P_2} \right) \qquad a_i = -10 \cdot \log \left(\frac{P_3 + P_4}{P_1} \right) \qquad (4.1)$$

directional coupling losses a_d: attenuation in "straightforward" direction:

$$a_d = -10 \cdot \log \left(\frac{P_1}{P_3} \right) \qquad (4.2)$$

coupling losses a_c: losses in "decoupling" direction:

$$a_c = -10 \cdot \log \left(\frac{P_1}{P_4} \right) \qquad (4.3)$$

return losses a_r: losses in "backward" direction:

$$a_r = -10 \cdot \log \left(\frac{P_1}{P_{1R}} \right) \qquad a_r = -10 \cdot \log \left(\frac{P_1}{P_{1R}} \right) \qquad (4.4)$$

coupling efficiency η:

$$\eta = \frac{P_3 + P_4}{P_1} \qquad (4.5)$$

splitting ratios S:

$$S = \frac{P_3}{P_4} \qquad (4.6)$$

crosstalk a_{ct} to port 2:

$$a_{ct} = -10 \cdot \log \frac{P_2}{P_1} \qquad (4.7)$$

Power P should be used in units of Watt (W) or milli Watt (mW), respectively.

Optical pathways are reversible. Thus, the description of couplers can also be used also for splitters. In Fig. 4.4 (see Sal 08), several couplers and splitters are sketched.

Most important couplers are fun-in and fun-out (for passive networks), 3 dB couplers and T-couplers (for splitting or combining channels) and star couplers.

4.2.2 Types of Couplers

Couplers are *passive* elements. In general, there are two variations of couplers:

- Front face couplers and
- Surface area couplers.

4.2.2.1 Face Couplers

In a T-splitter, incoming light power (P_1) will be distributed into two waveguides (P_2, P_3). It is used both in fiber technology (Fig. 4.5) and integrated optics technologies.

Fibers 2 and 3 should be ground with an angle of $\varphi = 1–2.5°$ to the axis.

Another possibility is the use of planar waveguides. In a glass-like material (e.g., silica glass) or crystalline material (e.g., lithium niobate $LiNbO_3$), the refractive index in the range where waveguide is planned can be increased by ion exchange. Examples are X- and Y-splitters in integrated-optics design (Fig. 4.6). The coupling ratio depends on the length of the range where waveguides are very close together (4.6b).

However, the losses that occur during the transition between glass fiber (cylindrical coordinates) and planar waveguide (Cartesian coordinates) and back must be considered in the calculation.

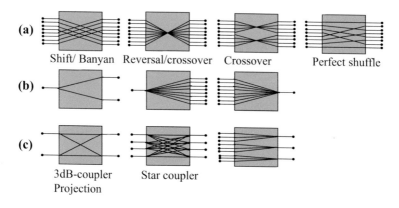

Fig. 4.4 Examples of couplers and splitters: One-to-one- (a), One-to-many- or Many-to-one- (b), and Many-to-many- interconnects (c)

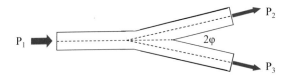

Fig. 4.5 Principle of T-splitter with fibers

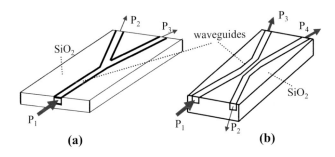

Fig. 4.6 Y-coupler (a) and X-coupler (b) in integrated-optical design

In *off-set splitters,* or TAP, surface areas are coupled directly (Fig. 4.7). In this case, fibers 2 and 3 have an offset with respect to the incoming fiber (Fig. 4.7a). This offset causes the coupling ratio. Due to high losses, this kind of coupler is not often used today. A mixing stick is less wasteful (Fig. 4.7b).

Coupling in a *lens splitter* is based on discrete elements like beam splitters and lenses (Fig. 4.8). Lens L_1 widens the beam from fiber 1. A beam splitter (BS) splits the beam into two parts. After that, the beams are focused by lenses L_2 and L_3 into fibers 2 and 3, respectively. These couplers work nearly lossless, but a great deal of alignment is required because incoming and outgoing fibers must be placed exactly in the focal points

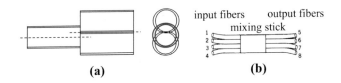

Fig. 4.7 Off-set splitter (a) and application of a mixing rod (b)

Fig. 4.8 Lens splitter

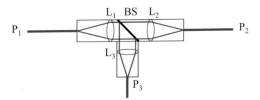

of the lenses. Furthermore, we have to consider the influence of polarization on beam splitting.

4.2.2.2 *Surface area couplers*
Tapers are optical adapters for tapered transition from one optical waveguide to another. In taper couplers, fibers should be separately heated and stretched. Thus, we get a lateral contraction. By sticking (gluing) or fusion, excellent optical contact can be achieved between two of these fibers. In this case, modes from the *core and cladding* will then be over-coupled into the other waveguide. The efficiency of the over-coupling depends on the length of the coupling range.

In Fig. 4.9, one can see a 1:1 splitting (so-called 3 dB coupler). Power at port 1 will be distributed homogenously to ports 3 and 4.

In *couplers with curved fibers,* coupling is achieved by accurately defined bending of fibers (Fig. 4.10). Coupling will be improved if one uses polished fibers. In this case, core parts of the fibers are touching, i.e., there are points of contact. The coupling efficiency depends on the length L of the "joint" core region.

For practical uses, *frequency-selective directional couplers* are of importance (Fig. 4.11a). If at port 1 we have two different wavelengths λ_1 and λ_2, at a certain coupling length L and a certain distance s of the two waveguides, the beam can be split into two parts of different wavelengths: λ_1 completely at port 3 and λ_2 completely at port 4. In this way, we are able to separate two data channels.

Fig. 4.9 Function of a 3 dB coupler

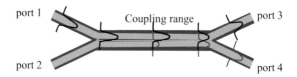

Fig. 4.10 Fiber coupler with polished core range

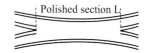

Fig. 4.11 Frequency-selective direction coupler (a) and transmitted power (b)

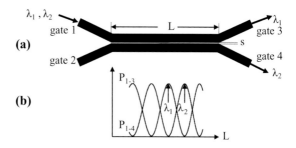

The wavelength dependence of transmitted power from port 1 to port 3 (P_{1-3}) or from port 1 to port 4 (P_{1-4}) is given by parameters L and s (Fig. 4.11b). For a certain value of coupling length L, waveguide distance s, and waveguide width, one gets a wavelength difference $\Delta\lambda = \lambda_2 - \lambda_1$, minimum value is about 13 nm. This limits the use of frequency-selected directional couplers as demultiplexers.

Exercise 4.12

What does 3 dB coupler mean?

▶ **Tip**
Help H4-12 (Sect. 12.1)
Solution S4-12 (Sect. 12.2)

Optical Transmitters

5

Today only semiconductor transmitters are used for optical transmission. Therefore, we will primarily focus on the basic setup of semiconductor light-emitting diodes (LED) and lasers.

5.1 Main Elements of Semiconductor Lasers

A laser consists of an energy source (pump), which has to transfer electrons in an active medium (AM) from a lower to a higher energy level (Fig. 5.1), e.g., from energy level E_1 (electron density $N_1 \gg 0$) to E_2 (electron density $N_2 \approx 0$).

A *spontaneous* emission of light results, with photon energy h·f exactly matched to the energy difference between active energy levels, i.e., $E_2 - E_1 = h \cdot f = \frac{h \cdot c_0}{\lambda}$. $h = 6.6 \cdot 10^{-34}$ Ws is Planck's constant, $c_0 = 2.99 \cdot 10^8$ m/s is the speed of light in vacuum, f is the (optical) frequency, and λ the wavelength. Spontaneous emission happens *randomly* in all directions and is of low intensity.

There are three conditions necessary for spontaneous to become *induced* or *stimulated* emission:

- There is an *inversion*, i.e., in level E_2 there are more electrons than in E_1 ($N_2 > N_1$).
- The higher laser level E_2 is *metastable*, i.e., electrons stay for a relatively long time (some milliseconds) on this level (without any spontaneous emission).
- There are some "starting" photons h·f of suitable wavelength $\left(\lambda = \frac{h \cdot c_0}{E_2 - E_1} \right)$, for example, photons from spontaneous emission.

Supplementary Information The online version contains supplementary material available at https://doi.org/10.1007/978-3-658-43242-3_5.

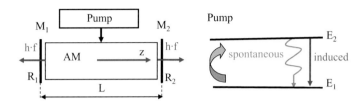

Fig. 5.1 Main elements of a laser and energy levels considered for laser process in an active medium (AM)

Stimulated emission generally goes in all directions, only by using feedback, e.g., through mirrors is the emission directed in a specific direction (z-direction in Fig. 5.1).

In fact, amplification results from induced emission, i.e., from one (e.g., spontaneously originated) photon, we get two photons of *identical frequency and phase*, later four, then eight, etc. To get an efficient amplification, a feedback is necessary. This can be realized by mirrors M_1 (reflection R_1) and M_2 (reflection R_2) of distance L. M_1 and M_2 form a *resonator* with the length L. A certain part of the light passes through the mirror and leaves the resonator—this light can be used for data transfer. The transmitted light is lost for the resonator. Another part of the light will be reflected and amplified by another induced emission in the active medium.

Amplification in propagation direction z can be described as

$$P(z) = P_0 e^{+gz} \tag{5.1}$$

g is the gain (unit: cm^{-1}) and P is the power (unit: mW).

Amplification is reduced by all losses in the resonator, i.e., losses in the active medium AM α (in cm^{-1}) and at mirrors (R_1 and R_2):

$$P(z) = P_0 \cdot e^{-\alpha z}; P_{M_1} = P_0 \cdot R_1; P_{M_2} = P_0 \cdot R_2 \tag{5.2}$$

P_{M_1} denotes power after mirror M_1 and P_{M_2}—power after mirror M_2.

During one complete resonator round trip (total length 2L), one gets the amplification condition:

$$P(2L) = P_0 \cdot e^{+2gL} \cdot e^{-2\alpha L} \cdot R_1 \cdot R_2 \tag{5.3}$$

To obtain an amplification after one round trip, the power must be more than a threshold value $P(2L) > P_0$. Thus, to get a positive total gain g, it must be more than a threshold gain g_{th}:

$$g \geq g_{th} = \alpha - \frac{1}{2L}(R_1 \cdot R_2) \tag{5.4}$$

Gain g can be enlarged by more pumping.

Exercise 5.1

The length of a resonator is (a) 1 mm and (b) 10 cm. Let the reflectivity of mirrors be 30%, the active medium is lossless. What is the minimum gain required for amplification to occur?

▶   **Tip**
    Help H5-1 (Sect. 12.1)
    Solution S5-1 (Sect. 12.2)

The following (more general) conditions should also be met for semiconductor lasers:

- The energy source (pump) of semiconductor lasers is electric current.
- The active medium is a semiconductor with a p-n transition, its composition should be chosen so that induced emission of light at the desired wavelength takes place.
- There are no extra mirrors in semiconductor lasers, one generally uses reflection—especially prepared end faces. In this case, we have the so-called Fresnel's reflection, (or as pdf file) where the refractive index of semiconductor ($n_{SC} = 3.2 \dots 3.6$) and air ($n_a = 1$) is important:

$$R = \frac{(n_{HL} - n_a)^2}{(n_{HL} + n_a)^2} \tag{5.5}$$

For example, with $n_{SC} = 3.2$ and $n_a = 1$, we can calculate the reflectivity as $R \approx 27.4\%$.

5.2   Active Element

The wavelength of the emitted light is given by semiconductor properties like band structure and energy gap [Blu 95].

5.2.1   Lattice Structure of Semiconductors

A crystalline lattice is necessary for semiconductor lasers. Of course, it is cheaper to use a disordered (amorphous) or a partially ordered (polycrystalline) semiconductor, but the energetic structure is too disordered.

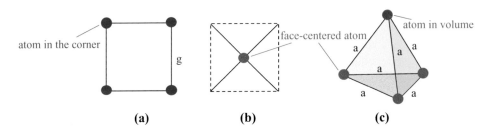

Fig. 5.2 Semiconductor structure: atoms in the corners (a), face-centered atoms (b), and atoms at the top of a tetrahedron (c)

Crystalline semiconductors consist of one or more atoms, e.g., silicon (Si), germanium (Ge), or gallium arsenide (GaAs). There are—for energy reasons—only two structures of the crystal lattice (Fig. 5.2):

- Diamond lattice for elemental semiconductor (consists only of atoms of the same type, e.g., Si or Ge).
- Zinc diaphragm lattice for compound semiconductor (consists of two different atoms, e.g., Ga and As).

Both lattices have the structure of a cube of edge length g of about 0.56 nm or (as usual in semiconductor physics) 5.6 Ångström (Å) with a total of 18 identical (e.g., Si atoms in the diamond lattice) or different atoms (e.g., smaller As and larger Ga atoms in the zinc blende lattice).

- Eight of them are arranged in the eight corners (Fig. 5.2a), e.g., Si or As.
- plus six face-centered atoms, each in the middle between four corner points (Fig. 5.2b), e.g., Si or As.
- the remaining four atoms are arranged at the tip of a congruent tetrahedron inside the structure, each in the center between three face-centered atoms and one atom in the corner point (Fig. 5.2c), e.g., Si or Ga.

Altogether, diamond and zinc blende lattices result as shown in Fig. 5.3. Details can be found in the "Extras" folder under "Semiconductor Structure.pdf".

To calculate the density of the atoms in the crystal lattice, one must consider the following:

- Each corner point is the common corner point of eight "cubes".
- The area of each face-centered atom is the common area of two "cubes".
- The four atoms in the volume occur only once in each "cube".

Thus, in each "cube" only one atom (from 8) "acts" in the corner point, three face-centered atoms (from 6), and four atoms in the volume. There are eight identical atoms per

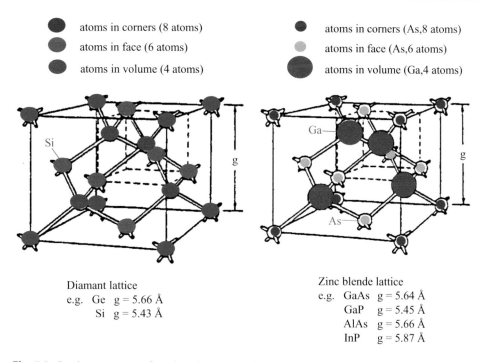

atoms in corners (8 atoms)   atoms in corners (As,8 atoms)
atoms in face (6 atoms)   atoms in face (As,6 atoms)
atoms in volume (4 atoms)   atoms in volume (Ga,4 atoms)

Diamant lattice   Zinc blende lattice
e.g. Ge g = 5.66 Å   e.g. GaAs g = 5.64 Å
      Si g = 5.43 Å   GaP   g = 5.45 Å
                                                AlAs g = 5.66 Å
                                                InP   g = 5.87 Å

Fig. 5.3 Lattice structures of semiconductor crystals

cube for element semiconductors (e.g., Si) and four atoms per component for compound semiconductors (e.g., four Ga atoms in corner points and face-centered, and four As atoms in the volume).

The particle density N_A then results as eight atoms per volume unit (g^3). This results in particle densities of about $4 - 5 \cdot 10^{22}$ cm^{-3}. The particle density sets the limit for the dopability, which we will need later.

Quantum mechanics calls such a cube a quantum dot or 0D (zero-dimensional) semiconductor. Although the semiconductor lattice consists of 18 atoms, it has an almost atomic energetic structure with very narrow energy levels (Fig. 5.4a). The energetic structure and therefore, emission of light of quantum dots (about 2 - 30 nm) is determined by their size. Quantum dots are currently the subject of intense research and are already being used in practical applications such as computer monitors, and TV screens based on quantum dot LED (QLED) technology. Note that the 2023 Nobel Prize in Chemistry (https://www.kva.se/app/uploads/2023/11/scibackkeen23.pdf) was awarded to Moungi G. Bawendi, Louis E. Brus and Alexei I. Ekimov for work in this field. If you string many quantum dots together, you get a quantum wire or a 1D (one-dimensional) semiconductor (Fig. 5.4b). A quantum well, a two-dimensional (2D) semiconductor structure, is formed from many parallel quantum wires (Fig. 5.4c). Only when many quantum films are superimposed does the long-known volume semiconductor with a 3D semiconductor structure emerge (Fig. 5.4d).

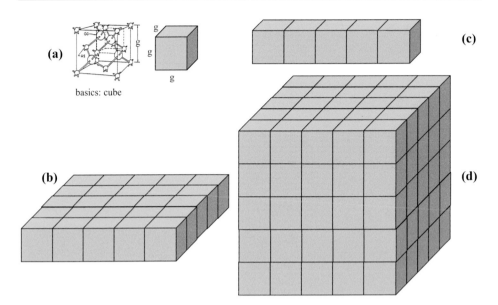

Fig. 5.4 From point-shaped semiconductor 0D (a) via semiconductor wire 1D (b), semiconductor films 2D (c) to volume structures 3D (d)

Fig. 5.5 Bands model of semiconductors

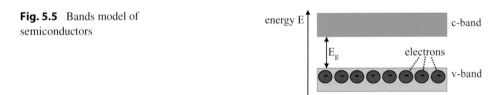

5.2.2   Band Structure of Semiconductors, Direct and Indirect Transitions

In contrast to discrete energy structure in atoms, semiconductors behave like crystalline solid states characterized by an energy band model (Fig. 5.5). From many different bands, the highest band which is *completely occupied* by electrons is the *valence band* or v-band, and the overlaying *completely empty* band is the *conductivity band* or c-band. An energy of electrons between c- and v-bands is forbidden, this is the forbidden zone or *band gap* E_g.

In general, light emission (spontaneous or induced) can occur only from the bottom edge of conductivity to the top edge of the valence band, respectively; thus $h \cdot f = \frac{h \cdot c_0}{\lambda} = E_g$. In reality, we have to consider not only energy but also the pulse (in a semiconductor the so-called 3D quasi-pulse $\vec{k}$ (Fig. 5.6). There are possible transitions without changes of quasi-pulse (*direct* semiconductor, e.g., GaAs) or with changes of

Fig. 5.6 Direct and indirect semiconductors

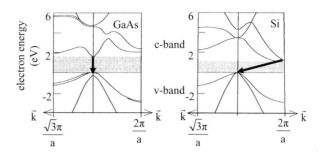

Table 5.1 Properties of selected elementary and binary semiconductors

Semiconductor	E_g (eV)	λ_g (μm)	direct/indirect	g (Å)
GaAs	1.424	0.871	direct	5.64
InP	1.351	0.918	direct	5.87
InAs	0.360	3.444	direct	6.14
InSb	0.172	7.208	direct	6.47
Si	1.22	1.107	indirect	5.43
Ge	0.66	1.878	indirect	5.66
GaP	2.261	0.548	indirect	5.43
AlAs	2.163	0.573	indirect	5.66

quasi-pulse (*indirect* semiconductor, e.g., Si or Ge). Only direct semiconductors can be used as transmitters.

In Table 5.1, one can find an overview on elementary (consisting of one element) and binary (composed of two elements) semiconductors. Using Planck's constant h and speed of light c_0, one gets the energy gap E_g. For the wavelength, one obtains Eq. (5.6), where precise units have to be used:

$$\lambda_g(\mu m) = \frac{1.24}{E_g(eV)} \tag{5.6}$$

Exercise 5.2

Why can indirect semiconductors not be used as transmitters?

▶ **Tip**
Help H5-2 (Sect. 12.1)
Solution S5-2 (Sect. 12.2)

Fig. 5.7 To the choice of
mixed crystals (ternary or
quarternary)

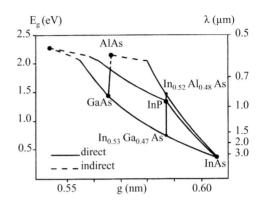

5.2.3 Choice of Material

From the table above, we see that we are *not* free to select a certain wavelength; thus, there is not just *one* material for *all* semiconductor lasers. For this reason, it is necessary—and possible—to combine semiconductors to *mixed crystals*. Thus, we get ternary (composed of three elements) or quaternary (composed of four elements) mixed crystals. One can mix either direct or indirect semiconductors. These mixed crystals composed of direct and indirect semiconductors can be used only in certain wavelength limits, where direct transitions are dominant (solid line in Fig. 5.7).

When mixing crystals, there is one distinct difficulty: The lattice constants are not "matched" (see Table 5.1). Therefore, no crystalline semiconductor structure occurs. The only exception is a mixture of GaAs with AlAs resulting in AlGaAs, but the available wavelength range (0.65–0.87 μm) is not useful for optical communication. Recently *tensile strained* or *compressive strained* semiconductors are being used (see Chap. 8—semiconductor amplifiers), but they are still in research. The problem of grating mismatch can be (commonly) by-passed, if on a certain substrate (platelets of GaAs or InP) ternary or quaternary semiconductors are epitaxially grown. For example, one can "mix" $In_{0.53}Ga_{0.47}As$ or $In_{0.52}Al_{0.48}As$ with InP. Thus, we can "create" laser light in wavelength range from 750 to 1700 nm.

Exercise 5.3

Why is it impossible to mix two different semiconductors, e.g., InAs and InP, to generate a suitable emission wavelength?

▶ **Tip**
   Help H5-3 (Sect. 12.1)
   Solution S5-3 (Sect. 12.2)

The basic idea can be illustrated by some "cooking recipes" [Har 98]. The formulas mentioned mainly result from the requirement to match the grating constants g. In these formulas, wavelength λ must be given in μm.

The composition of ternary mixed semiconductors for the first optical window ($\lambda \approx 800$ nm):

System: $Ga_{1-x}Al_x$ as on GaAs

GaAs (substrate)

Band gap $E_g = 1.424$ eV, thus $\lambda = 1.24/E_g = 0.871$ μm

Lattice constant $g = 5.653$ Å

Refractive index $n = 3.59$ (at $\lambda = 0.9$ μm)

$Ga_{1-x}Al_x$ As (matched grating to GaAs)

Band gap $E_g = (1.424 + 1.247 \cdot x)$ eV ($0 < x < 0.45$)

Lattice constant $g = (5.653 + 0.027 \cdot x)$ Å

Refractive index $n = 3.59 - 0.71 \cdot x$ (at $\lambda = 0.9$ μm)

Exercise 5.4

Calculate the exact composition of a GaAlAs mixed crystal for $\lambda = 850$ nm!

▶ **Tip**
Help H5-4 (Sect. 12.1)
Solution S5-4 (Sect. 12.2)

The composition of quaternary mixed semiconductors for the second and third optical windows (wavelength from 1300 to 1550 nm):

System: $In_{1-x}Ga_x As_y P_{1-y}$ on InP

InP (substrate)

Band gap $E_g = 1.35$ eV, thus $\lambda = 0.9185$ μm

Lattice constant $g = 5.8696$ Å

Refractive index $n = 3.21$ (at $\lambda = 1.3$ μm)

$In_{1-x}Ga_x As_y P_{1-y}$ (lattice matched to InP)

Band gap $E_g = (1.35 - 0.72 \cdot y + 0.12 \cdot y^2)$ eV $y = 3 - \sqrt{\frac{1.24 - 0.27 \cdot \lambda}{0.12 \cdot \lambda}}$

             for $0 < x < 0.47$ and $0 < y < 1$

             $x = y/(2.2091 - 0.06864 \cdot y))$

Lattice constant $g = (5.8696 + 0.1894 \cdot y - 0.4148 \cdot x + 0.013 \cdot x \cdot y)$ Å

Refractive index $n = 3.4 + 0.256 \cdot y - 0.095 \cdot y^2$ (at $\lambda = 1.3$ μm)

Exercise 5.5

Calculate the exact composition of a GaAlAs mixed crystal for $\lambda = 1300$ nm!

▶ **Tip**
Help H5-5 (Sect. 12.1)
Solution S5-5 (Sect. 12.2)

Exercise 5.6

Calculate the exact composition of a GaAlAs mixed crystal for $\lambda = 1550$ nm!

▶ **Tip**
Help H5-6 (Sect. 12.1)
Solution S5-6 (Sect. 12.2)

The composition of quaternary mixed semiconductors for the second and third optical windows (wavelength from 1300 to 1550 nm):

System: $In_{1-x-y} Ga_x Al_y$ as on InP

InP (Träger)

Band gap $E_g = 1.35$ eV, thus $\lambda = 0.9185$ μm

Lattice constant $g = 5.8696$ Å

Refractive index $n = 3.21$ (at $\lambda = 1.3$ μm)

$In_{1-x-y} Ga_x Al_y$ As (grating matched to InP)

Band gap $E_g = (0.75 + 1.0496 \cdot x + 1.0645 \cdot x^2 - 0.033 \cdot x^3)$ eV

$\quad\quad\quad y = 0.468 - 0.983x$

$\quad\quad\quad$ for $0 < x < 0.476$; $0 < y < 0.468$

Refractive index $n = 3.595 - 1.103y + 0.745y^2$ (at 1.55 μm)

Exercise 5.7

Calculate the exact composition of a GaAlAs mixed crystal for $\lambda = 1300$ nm!

▶ **Tip**
Help H5-7 (Sect. 12.1)
Solution S5-7 (Sect. 12.2)

Exercise 5.8

Which composition could a mixed semiconductor transmitter have for $\lambda = 1550$ nm?

▶ **Tip**
Help H5-8 (Sect. 12.1)
Solution S5-8 (Sect. 12.2)

Exercise 5.9

We need two transmitters with wavelengths close by (1522 and 1523 nm). Which active material do you propose?

▶ **Tip**
Help H5-9 (Sect. 12.1)
Solution S5-9 (Sect. 12.2)

Exercise 5.10

Why do we use InP and not GaAs as substrate for quaternary mixed semiconductors?

▶ **Tip**
Help H5-10 (Sect. 12.1)
Solution S5-10 (Sect. 12.2)

5.2.4 Light Emission in Semiconductors, LED

5.2.4.1 Recombination in Semiconductors

By pumping it is possible to bring many electrons to the c-band. Because these electrons are now in a higher energetic state, they are set to unload (get rid of) this energy. There are basically three possibilities:

- linear recombination (see also pdf file)
  In each semiconductor, we can find a small number of impurities or inhomogeneities in crystal structure resulting in energy levels close to the c-band edge (dotted line in Fig. 5.8a). These energy levels act as trap centers for electrons in the conductivity band (their number should be N_T). By these trap centers, the electron density in the c-band N_c is decreased exponentially in time, see (5.7). These electrons are lost for light emission.

$$\frac{dN_c}{dt} = -\text{const.} \cdot N_c \quad \text{solution}: N_c(t) = N_c(0) \cdot e^{-\frac{t}{\tau_L}} \tag{5.7}$$

At time τ_L, number N_c drops down to the 1/e part. This time τ_L is denoted as the lifetime. Typical values for τ_L are about 10 ms (in crystalline semiconductors) down to 10 ps (in amorphous semiconductors). For this reason, amorphous semiconductors cannot be used for transmitters.

- quadratic or bimolecular recombination (see also pdf-File)

  At quadratic or bimolecular recombination, an electron "jumps" from c-band directly into a hole in v-band (Fig. 5.8b). In this case, energy $hf = E_g$ in terms of light emission releases. This photon energy corresponds to the energy gap between c- and v-bands. In this process, two types of particles are involved (N_c electron and P_v hole)—this is why one can use the term "bimolecular". For temporal decay of electron density, we get the equation below (Eq. 3.1). In general, electrons and holes have the same density ($N_c = P_v$).

$$\frac{dN_c}{dt} = -\text{const.} \cdot N_c \cdot P_v \approx \text{const.} \cdot N_c^2 \quad \text{solution}: N_c(t) = \frac{N_{c_0}}{1 + B \cdot N_{c_0} \cdot t} \tag{5.8}$$

The characteristic decay time is $\tau_{\text{quadr}} = \frac{1}{B \cdot N_{c_0}}$. Typical values for crystalline semiconductors are $B = (0.3 \dots 2) \cdot 10^{-10} \text{cm}^3/\text{s}$. These decay times are in the ns range. Lifetime τ_L is connected with spontaneous emission of light—this is the light-emitting diode (LED). Only quadratic recombination with decay time τ_{quadr} can be used for laser processes. This is why the electron density chosen should be as high as possible—the higher the electron density, the more the quadratic recombination and the emitted light power.

Exercise 5.11

Which decay time does one get for electrons at $N_{c_0} = 10^{17} \text{ cm}^{-3}$ or 10^{18} cm^{-3} and $B = (0.3 \dots 2) \cdot 10^{-10} \text{cm}^3/\text{s}$?

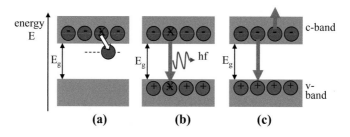

Fig. 5.8 Linear (a), quadratic (b), and Auger recombination (c)

▷ **Tip**
Help H5-11 (Sect. 12.1)
Solution S5-11 (Sect. 12.2)

- Auger recombination (see also pdf file)
  At very high densities N_c and P_v, the "jump" of electrons into holes (like at quad-ratic recombination) is not combined with emission of light but with a transfer of this energy to another particle, e.g., to an electron in c-band (Fig. 5.8c). This *radiation-less* recombination first was described by Pierre Victor Auger (1899–1993) and named Auger recombination. Three particles take part in this process—electrons N_c and holes P_v; we get Eq. (5.9). In general electrons and holes have again the same density $(N_c = P_v)$. Characteristic decay time τ_{Auger} depends quadratically on the electron and hole density—the higher the density, the faster (and more effective) the Auger recom-bination.

$$\frac{dN_c}{dt} = -\text{const.} \cdot N_c \cdot (P_v + N_c)\tau_{\text{Auger}} = \frac{1}{4C\left(N_{c_0} + P_{v_0}\right)^2} = \frac{1}{4C \cdot N_{c_0}^2} \qquad (5.9)$$

- Typical values for crystalline semiconductors are $C = (1 \dots 5) \ 10^{-29}$ cm^6 s^{-1}. Thus, at very high electron and hole densities, we get extremely short decay times (in the ps range). For laser operation, this means that electron and hole densities which are too high are destructive and should be avoided.

Exercise 5.12

Which decay time does one get for electron densities of $N_{c_0} = 10^{19}$ cm^{-3} or $N_{c_0} = 10^{20}$ cm^{-3} with $C = (1 \dots 5) \cdot 10^{-29}$ cm^6/s?

▷ **Tip**
Help H5-12 (Sect. 12.1)
Solution S5-12 (Sect. 12.2)

Summarizing we can say that, for laser operation, only quadratic recombina-tion is important. Thus, for adequate electron and hole densities, we get a window $(<10^{19}$ cm$^{-3})$, where Auger recombination is not significant.

5.2.4.2 Line Width

In accordance with the energy-band model (Fig. 5.8), quadratic recombination results in emission of exactly *one* wavelength or frequency $hf = E_g$. In reality, we get a narrow spectrum with the fluorescence or emission line width $\Delta\lambda_F$ or Δf_F, respectively. The ori-gin of this spectrum can be explained in the following way:

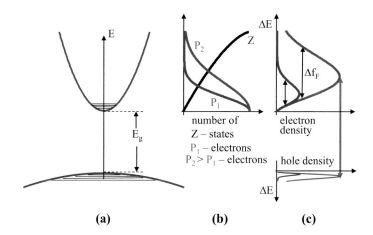

Fig. 5.9 Origin of line width: Band structure (a), state density and electron number (b), and electron and hole density (c)

- There are a certain number of "locations" in c- and v-bands (states Z in Fig. 5.9a). The number of these states increases with the distance to the band edge ΔE (Fig. 5.9b): $Z \propto E^{3/2}$.
- These states will be occupied with electrons (P_1, P_2); as a model, one can use the Gaussian distribution of electrons (see also pdf file) of Fig. 5.9b: $P \propto e^{-\Delta E^2}$. $P_2 > P_1$ means higher pumping (higher electric current).
- Convolution of Z and P gives the electron density (Fig. 5.9c). The half-width of electron density is then $\Delta\lambda_F$ or Δf_F (Fig. 5.9c). The same consideration can be done for holes.

Quadratic recombination means the "jump" of an electron of c-band into a hole of v-band. Because starting and end points can vary, thus we get the half-width of emission $\Delta\lambda_F$ or Δf_F. Typical values are $\Delta\lambda_F = 30 - 60$ nm. From Fig. 5.9c, we can draw two conclusions:

- The higher the pumping the lower the central wavelength.
- The higher the pumping the broader the line width.

Light emission with a line width $\Delta\lambda_F$ represents the typical spectrum of an LED.

5.2.4.3 p-n Junction as a Basic Structure, LED
For quadratic recombination at the same place, a large number of electrons in c-band and holes in v-band are necessary. This can be achieved with a p-n transition. Normally semiconductors can be considered as intrinsic (i-semiconductor). Conductivity of semiconduc-

Fig. 5.10 p-n junction as basic structure, LED

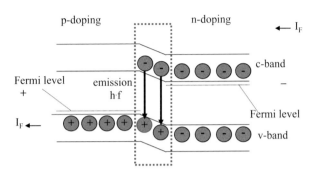

tors can be changed by doping with acceptors (e.g., p-GaAs with doping of Zn, Cd, or Si instead of Ga) or donors (e.g., n-GaAs with doping of S, Se, Te, or Si instead of As).

The basic structure of a semiconductor laser is depicted in Fig. 5.10. Electron distribution in permitted states—here in v- and c-bands—plays an important role in semiconductors. The virtual energy level, where the probability to find an electron is 50%, is the so-called Fermi level (calculations as Fermi level, see also pdf-File). In non-doped semiconductors, we have no permitted states in the forbidden zone (energy gap)—the energy level with 50% occupation probability is exactly in the middle of the energy gap. The n-doping results in permitted states near to the c-band—the Fermi level is shifted into c-band direction. Vice versa at p-doping, we find permitted states near to the v-band— the Fermi level is shifted into v-band direction (see Fig. 5.10). The p-n junction consists of a contact between p-doped and n-doped semiconductors. Electric current (I_F) has to run from minus to plus [Wag 98]. The potential difference results in a situation, where, in the range of p-n junction, we have electrons in c-band and holes (= "missing" electrons) in v-band nearby—spontaneous or induced emission by quadratic recombination can start. The mobility of electrons is much higher than that of holes. For example, electron mobility in GaAs is $\mu_e = 8500$ cm^2/Vs and mobility of holes is only $\mu_h = 435$ cm^2/Vs. Therefore, quadratic recombination takes place much more often in the p-doped part of the p-n junction. The p-n junction is the basic element of LED's as well as of semiconductor lasers.

5.2.5 Semiconductor Transmitters—Basic Structure

5.2.5.1 *Double-Hetero-Structure Laser (DH Diode)*

If the p- and n-doped semiconductors are of the same material (e.g., GaAs or a mixed crystal InGaAsP for suitable wavelength), then we have the so-called Homo-Diode (HD). In Homo-Diodes, the recombination zone is restricted to a few μm in the p-range. Therefore, a high current density is necessary for laser operation, and an operation at room temperature (without cooling) seems to be impossible.

Fig. 5.11 Emission, refractive index, and energy-band structure of a DH semiconductor lasers

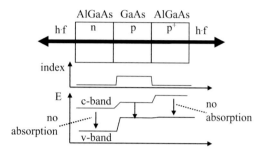

This problem can be solved by use of Double-Hetero-structures (Fig. 5.11). Today this is the most used semiconductor laser. The p-range necessary for quadratic recombination consists of the semiconductor *of chosen wavelength* (e.g., GaAs or InGaAsP). It is surrounded by another semiconductor *with higher energy gap* (e.g., AlGaAs for GaAs laser). Thus, emitted light can propagate in the AlGaAs-layer without absorption, and we can avoid any *re-absorption*. This situation is depicted in Fig. 5.11 by the length of arrows with respect to the energy gap. If the p-range consists of one semiconductor, and the surrounding consists of an n− or p⁺-ranges (p⁺ means simply higher doping) we have the situation known as a Double-Hetero (DH) structure.

A welcome side effect is that the refractive index in the p-range is higher than in the other ranges (Fig. 5.11)—similar to glass fibers, we then get waveguiding in the p-doped range. Thickness of this laser-active p-range can now be reduced down to less than 1 μm (practically down to 0.2 μm or less). This is enough to get laser operation at relatively low current.

Exercise 5.13

Why does laser operation depend significantly on the thickness of the p-doped semiconductor range?

▶   **Tip**
    Help H5-13 (Sect. 12.1)
    Solution S5-13 (Sect. 12.2)

5.2.5.2 From the Volume Semiconductor to the Point Semiconductor

Previous considerations were based on volume semiconductor structures with an edge length of about 100 μm. Characteristic for such semiconductors is a band structure with valence and conduction band.

5.2.5.3 *Multi-Quantum Well Structure (MQW)*

If one decreases the thickness of the laser-active layer (e.g., GaAs) down to atomic dimensions (down to 1–2 nm, i.e., comparable to the grating constant, see Table 5.1) a band structure will no longer exist. In this case, we have discrete energy levels such as in

Fig. 5.12 MQW structure

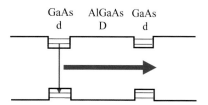

atoms. The position of these energy levels depends on the thickness of layer d and of the surrounding semiconductor. Like in DH structures, this surrounding consists of another semiconductor with another (higher) energy gap and with thickness D. Thus, a quantum well is formed in which the electrons are trapped. If we have many quantum wells, we get a Multi-Quantum Well (MQW) structure (Fig. 5.12).

A typical MQW laser consists of approx. 10–15 quantum wells with a "thickness" of $d = 1 - 20$ nm and $D = 5 - 50$ nm, respectively. The high number of quantum wells is necessary to achieve high total amplification and therefore high laser power.

The main advantages of MQW lasers are:

- Low layer thickness results in a reduced threshold current.
- Temperature dependence of the threshold current is low.
- Depending on thickness d and ratio d/D, we can generate different wavelengths (at GaAs MQW laser down to about 700 nm. This method of controlling the laser wavelength is called Band Gap Engineering.
- Due to narrow line width of discrete energy levels, MQW lasers have only one longitudinal mode and an extremely narrow linewidth (especially as DFB laser, see Sect. 5.3.2).
- MQW lasers can operate continuously (continuous waves or cw laser), up to 100 mW is possible.
- Lifetime is high (about 10^5 h corresponding to 12 years of continuous operation).

Exercise 5.14

A Fabry-Perot laser consists of a quaternary mixed semiconductor and emits a wavelength of 1520 nm. In which direction will the wavelength be shifted if one uses an MQW laser with the same mixed semiconductor?

▶   **Tip**
Help H5-14 (Sect. 12.1)
Solution S5-14 (Sect. 12.2)

5.3   Resonators

Feedback necessary for laser operation can be achieved by different methods:

- One can use compact mirrors with a reflectivity R. For semiconductor lasers, these are the end faces with a reflectivity of about 30%. Reflectivity can be calculated by Fresnel's Eq. (5.5). With the refractive index of a semiconductor ($n_{sc} \cong 3.4$) and air ($n_{air} = 1$), we can calculate R $= 30\%$. The corresponding laser type is a *Fabry-Perot laser*. The majority of existing semiconductor lasers (e.g., in CD players, CD burners etc.) are Fabry-Perot lasers.
- Mirrors are "distributed" along the active material. One can imagine it as a high number of mirrors with very small reflectivity. This type of laser is of advantage for optical data transmission with very high bit rates.

5.3.1   Fabry-Perot Laser

A Fabry-Perot laser uses compact mirrors; in semiconductor lasers, the reflection is at the end faces. With a DH structure, we get a setup as depicted in Fig. 5.13 with AZ—laser active zone.

A Fabry-Perot laser is low cost, otherwise, due to its spectral properties; this laser can only be used for optical communication in certain cases. Because only standing waves between mirrors can interfere constructively, some additional wavelengths result—the so-called *longitudinal mode*—besides the main wavelength. These equidistant, longitudinal modes have the tendency for a cross-talk to the neighboring channel.

The origin of longitudinal modes in a resonator of *optical length* $L = L_0 n$ can be explained using the wave model to calculate the dependence of amplitude transmission

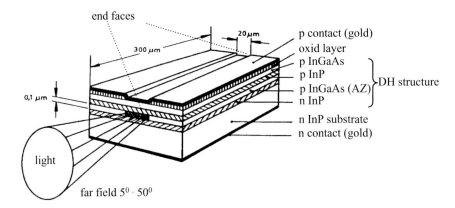

Fig. 5.13 Fabry-Perot laser with DH structure

Fig. 5.14 To the origin of longitudinal modes

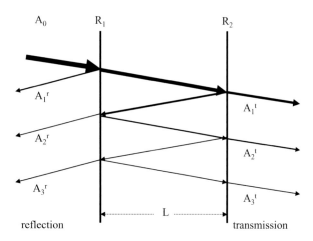

on the wavelength (Fig. 5.14). Transmitted amplitudes A_1^t, A_2^t, A_3^t etc., interfere, considering pathway differences (phase differences) at forward and backward propagation between mirrors with the reflectivity R_1 and R_2. Superposition of all amplitudes results in a geometrical row. And, in the case of $R_1 = R_2 = R$, the so-called Airy formula (transmission formula) results:

$$T(\lambda) = \frac{1}{1 + \frac{4R}{(1-R)^2} \cdot \sin^2 \frac{\delta}{2}} \quad \text{with } \delta = \frac{4\pi L}{\lambda} \tag{5.10}$$

We can get a similar formula for the reflected amplitudes A_1^r, A_2^r, A_3^r etc., but it seems simpler to calculate the reflection by formula $R(\lambda) = 1 - T(\lambda)$.

The evaluation of the Airy formula, e.g., using MathCad with $L = 1$ mm and $R = 0.35$, yields equidistant maxima of the transmission at the frequency spacing $\Delta \nu = c/2L = 1.5 \cdot 10^{11}$ Hz or a the wavelength spacing of $\Delta \lambda = \lambda_0^2/2L = 1.2$ nm at $\lambda_0 = 1.55$ μm (Fig. 5.15a). A pdf version can also be found.

If this transmission spectrum is superimposed on the fluorescence or emission spectrum of the semiconductor (assumed to be Gaussian, see Fig. 5.22, with the half-width $\Delta \lambda_F = 42$ nm), the longitudinal mode image is obtained as shown in Fig. 5.15b (also as pdf file).

Within the emission line width of $\Delta \lambda_F = 42$ nm, approx. 34 longitudinal modes occur, each of which has a line width of less than 1 nm.

Exercise 5.15

How much is the frequency distance between neighboring longitudinal modes in a Fabry-Perot laser, if the length of the optical resonator is 0.5 mm?

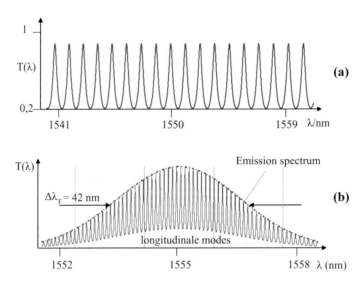

Fig. 5.15 Transmission spectrum of a resonator (a) and of a Fabry-Perot laser (b) with length L=1 mm

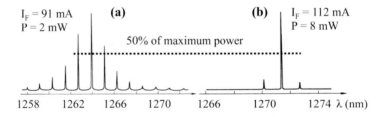

Fig. 5.16 Example of a Fabry-Perot laser spectrum (after [Ker 83])

▶ **Tip**
Help H5-15 (Sect. 12.1)
Solution S5-15 (Sect. 12.2)

In Fig. 5.16, the real modal spectrum of a Fabry-Perot semiconductor laser is depicted. From the distance between the modes of about 1 nm, one can conclude the resonator length L=1.1 mm. From Fig. 5.16a, one can see that power is concentrated mainly within three modes with a power more than 50% of the maximum value. Both the current optical power as well as modal picture are changed (Fig. 5.16b). Thus, under certain circumstances, one can consider a Fabry-Perot laser as a longitudinal single-mode laser. Otherwise, we cannot expect that this modal spectrum will remain unchanged with increasing power.

Exercise 5.16

How many longitudinal modes can we have in a Fabry-Perot laser with a total emission width of 50 nm and a resonator length of the active medium of 0.5 mm ($\lambda = 1550$ nm)?

▶ **Tip**
Help H5-16 (Sect. 12.1)
Solution S5-16 (Sect. 12.2)

5.3.2 Dynamical Single-Mode Laser (DSM)

In a *dynamical single-mode laser*, mirrors are no longer concentrated, but "distributed" along the active medium (Fig. 5.17). Like in a grating structure, refractive index alternates with a distance $\lambda/2n$. Higher (n_H) and lower (n_L) refractive indices can be achieved by ion bombardment. In contrast to conventional line gratings where a line will be scratched into the surface, changes in refractive index in DSM lasers occur in the complete volume of the active medium. The difference in refractive index is typically $n_H - n_L = 10^{-2} - 10^{-3}$, thus the reflectivity of a single "mirror" is about $0.2 \cdot 10^{-5} - 0.2 \cdot 10^{-7}$. To get high total reflectivity, one needs many "mini mirrors".

Another problem arises from a theoretical consideration of transmission in a system with *coupled* mirrors with distance D: Just at the desired emission wavelength, the transmission is zero at $D = \lambda/2n$. Therefore, we have to "add" an additional phase shift. This can be done by alteration of the grating period to D/2 or 3D/2. In doing so, it is irrele-

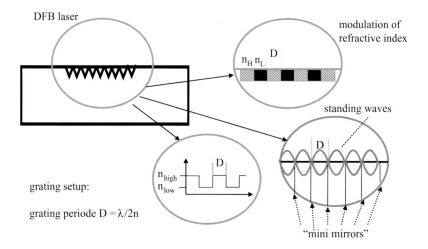

Fig. 5.17 Distributed feedback in DSM lasers

Fig. 5.18 DFB and DBR laser

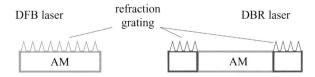

vant whether the changed period is realized "in one piece" or distributed along the active medium.

Exercise 5.17

Which reflectivity do we get in a DSM laser at a high-low change-over (mini mirror) if the difference between refractive index is 10^{-2}?

▶ **Tip**
Help H5-17 (Sect. 12.1)
Solution S5-17 (Sect. 12.2)

There are two possibilities for configuration of a DSM laser (Fig. 5.18):

- at DFB laser (Distributed FeedBack-Laser) the gratings are distributed along the active medium (AM).
- at DBR laser (Distributed Bragg Reflection) gratings are arranged outside of the active medium.

The spectral ratio in a DSM laser, i.e., wavelength dependence of the transmission, can be calculated by replacing the resonator length L in Eq. (5.10) with the grating period D:

$$T(\lambda) = \frac{1}{1 + \frac{4R}{(1-R)^2} \cdot \sin^2 \frac{\delta}{2}} \quad \text{with } \delta = \frac{4\pi D}{\lambda} \tag{5.11}$$

Because, in these "mini" resonators only wavelength $\lambda/2 = D$ is suitable to form a standing wave, we get emission of a single narrow laser line (Fig. 5.19). Note that wavelength inside the medium is changed to $\lambda = \frac{\lambda_0}{n_{HL}}$. With about 1000 round trips in the laser resonator with distributed feedback, we obtain a line width of about $\Delta\lambda = 10^{-2} \dots 10^{-4}$ nm.

Exercise 5.18

Describe the spectra of LED, Fabry-Perot, and DSM lasers. What is the reason for these differences?

▶ **Tip**
Help H5-18 (Sect. 12.1)
Solution S5-18 (Sect. 12.2)

Fig. 5.19 Spectrum of a DSM laser and emission spectrum of an LED

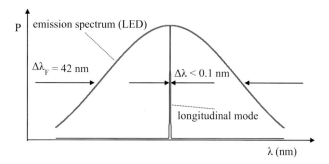

5.4   Laser Properties

In general, laser properties are based on the active semiconductor medium (induced emission, emission line width) and resonator properties. Let us describe the basic laser properties.

5.4.1   P-I Characteristic, Temperature Behavior, Degradation

The dependence of optical power p on pump current I_F in flow direction is given by the P-I or laser characteristic. In the laser characteristic (Figure 5.20), one can find a threshold current I_{th}. Below that threshold, spontaneous emission dominates (LED behavior); above the threshold, induced emission dominates—this is the actual "laser".

Furthermore, we have to consider the temperature dependence of P-I characteristic (Fig. 5.21a). With increasing temperature, the threshold I_{th} increases too, whereas the slope of P-I curve is changed only marginally. Only lasers with higher wavelength show a reduced slope with increasing temperature (dashed line in Fig. 5.21a). At constant pump current with increasing temperature, the power drops—thus constant operation temperature is required.

The aging process (degradation) of semiconductor lasers (Fig. 5.21b) runs in the same way. Besides an increase of the threshold current I_{th} the slope of the P-I curve—to get the same power more current is necessary. Per definition, the *lifetime* of a semiconductor laser is achieved if, at constant current, the power is reduced to half of the initial value (mostly many thousands of hours).

Exercise 5.19

What is the difference between the P-I characteristic of LED and semiconductor lasers?

Fig. 5.20 P-I characteristic

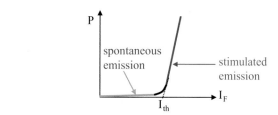

Fig. 5.21 Influence of temperature (a) and aging (b) to the P-I characteristic

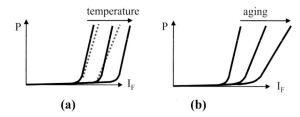

▶ **Tip**
Help H5-19 (Sect. 12.1)
Solution S5-19 (Sect. 12.2)

Exercise 5.20

How is power changed with increasing temperature and/or aging?

▶ **Tip**
Help H5-20 (Sect. 12.1)
Solution S5-20 (Sect. 12.2)

5.4.2   Spectrum of Semiconductor Lasers

For more practical operation of semiconductor lasers, the central (preferably constant) wavelength λ_0 and the line width $\Delta\lambda$ are of importance. Line width (see Sect. 5.2.4.2) is typically measured as half-width: the wavelength where power is reduced to 50%. Often one can find the abbreviation FWHM (Full Width at Half Maximum). It is very important that no spectral portions exist outside of the spectrum depicted in Fig. 5.22. Otherwise, a "cross-talk" can occur.

Fig. 5.22 Emission spectrum
of lasers

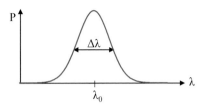

Fig. 5.23 Directional
characteristic of surface-
emitting (a) and edge-emitting
diodes (b)

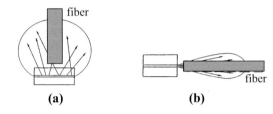

(a)   (b)

5.4.3   Radiation Characteristics

Electromagnetic radiation of light in different directions can be described by the *directional characteristic*. The directional characteristic is given by construction of the active medium and by the resonator. In the case of a *surface-emitting diode* (or so-called *Lambert radiator*), one gets a cone-shaped directional characteristic (Fig. 5.23a), generated from an area of about 50 μm diameter. An *edge-emitting diode* radiates from a surface area of circa 2 μm · 0.2 μm, and one gets an elliptical beam cross section (Fig. 5.23b). The radiation characteristic has consequences for the coupling of the light into the optical fiber since laser light with a circular beam cross section can be focused much more easily.

Exercise 5.21

What is the difference in directional characteristics between surface-emitting and edge-emitting diodes? Which directional characteristic is better matched to fiber systems and why?

▶   **Tip**
Help H5-21 (Sect. 12.1)
Solution S5-21 (Sect. 12.2)

5.5   Selected Laser Types for Optical Networks

For data transfer in optical networks, transmitters (semiconductor lasers) should meet certain requirements:

- They should be tunable, mainly in the range of 3rd optical window (1490–1560 nm).
- They should have a narrow bandwidth, especially for dense wavelength division multiplexing (DWDM); line width less than 10^{-4} nm is required.
- Wavelength should be very stable with respect to temperature and current changes.
- They should have high power (about 10 dBm) to cover a long distance without amplification.

In general, these requirements can be met only by MQW lasers with DSM resonator structure.

5.5.1 MQW Laser with DFB Resonator as Edge-Emitting Diode

Recent extremely narrow-band lasers use MQW laser structure with distributed feedback (DFB) as resonator. As described in Sect. 5.3.2, one gets at a certain wavelength, i.e., with a grating period $D = \lambda/2$, a transmission of 100%—however, reflectivity at the desired wavelength λ_0 becomes $R = 0\%$ and no feedback and no laser operation takes place. A way out of this dilemma can be achieved with a so-called *chirped* laser resonator. Expression "chirp" is from English, it means that when birds "chirp" the frequency increases. In this chirped resonator, the grating period is no longer constant, but changed either stepwise or continuously in such a way that the total phase shift along the resonator is $(2n+1).\pi$, where n is a whole number $(n = 0, 1, \ldots)$. Such a chirped grating structure with an additional phase shift of π is depicted in Fig. 5.24.

The setting up of a chirped grating as depicted in Fig. 5.24 is very complicated. It seems to be simpler to concentrate the additional phase shifts to several regions—at these regions, the grating period should be increased in total to 3D/2 or decreased to D/2. Such a version is shown in Fig. 5.25 [Hil 95]. Note that the power supply is split into the chirped and unchirped regions, respectively. As one can see, the grating period is increasing from L_1 to L_{p+1}.

A more elegant version of "installation" of phase shift was developed by Deutsche Telekom [Hil 95]. In this case, the MQW structure with DFB grating is not straightlines but curved. Bending should be calculated so that one gets again an additional phase shift of π (Fig. 5.26).

In Fig. 5.27, the structure of an MQW laser with bent grating is depicted. Besides gold and gold-germanium contacts, the MQW laser consists of InGaAs layers (active

Fig. 5.24 Comparison between unchirped and chirped grating

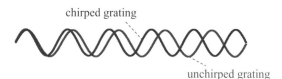

chirped grating

unchirped grating

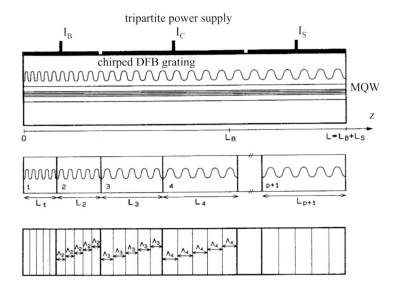

Fig. 5.25 Chirped grating

Fig. 5.26 Curved grating

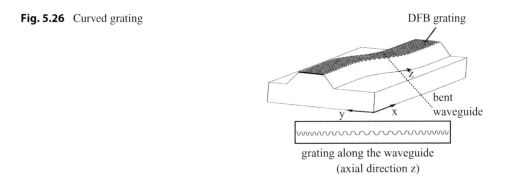

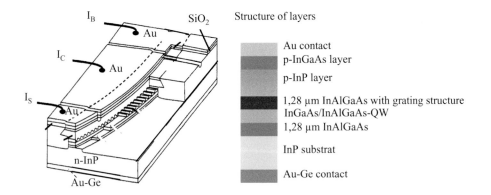

Fig. 5.27 Structure and layers of a MQW laser with bent grating

Fig. 5.28 Tunable MQW
laser with bent DFB grating
structure

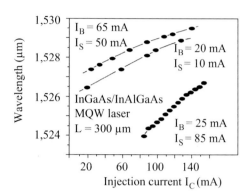

region) and InAlGaAs layers embedded into an InP substrate or InP layer, respectively. Grating is integrated into a 1.28 μm layer of InAlGaAs.

This laser can be tuned in wavelength within about 6 nm (Fig. 5.28). Tuning will be controlled by different currents I_B, I_S, or I_C, respectively. As with all edge radiators, coupling into a single-mode optical fiber is problematic due to the radiation characteristics.

Exercise 5.22

Which possibilities for wavelength tuning in a MQW-DSM laser do you know?

▶   **Tip**
   Help H5-22 (Sect. 12.1)
   Solution S5-22 (Sect. 12.2)

5.5.2   Vertical Cavity Surface Emitting Laser (VCSEL) as Surface-Emitting Diode

In surface-emitting diodes, light is emitted from an area with a diameter of about 10 μm. Thus, we get a spherical directional characteristic, which is much easier to focus into an SMF of 9 μm standard diameter than is possible with edge-emitting diodes. Note that the combination of an MQW with a DBR mirror permits single-mode operation (Fig. 5.29).

Details of VCSELs can be found on internet; they are available in all wavelengths between 800 and 1700 nm. Power of a few hundred mW is available.

Different directional characteristics of VCSEL, LED, and edge-emitting diodes are depicted in Fig. 5.30. VCSELs can be arranged in arrays without any problems. Thus, output power can be increased at reasonable focusability of the beam.

VCSELs have several advantages:

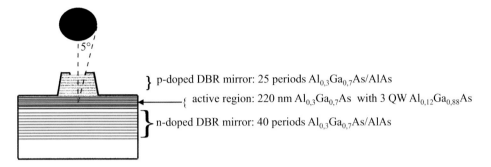

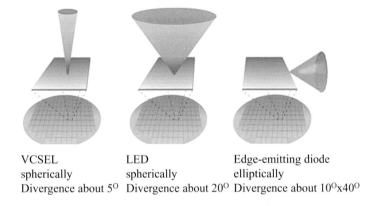

Fig. 5.29 Basic Setup of a Vertical Cavity Surface Emitting Laser

VCSEL   LED   Edge-emitting diode
spherically   spherically   elliptically
Divergence about 5° Divergence about 20° Divergence about 10^Ox40^O

Fig. 5.30 Directional characteristics of VCSEL, LED, and edge-emitting diodes

- Wavelength stability; operation in a single longitudinal mode.
- Wavelength uniformity & spectral width. Thus, fabrication of VCSEL 2D arrays with little wavelength variation between the elements of the array (<1 nm) is possible.
- Temperature sensitivity of wavelength (~5 times less sensitive to temperature variations than in edge emitters).
- High-temperature operation (uncooled operation up to 80 °C).
- Higher power per unit area (about 500W/cm^2, up to 2–4 kW/ cm^{2}can be expected in the near future.
- Higher beam quality due to emission of a circular beam.
- Reliability: typical failures in one billion device-hours for VCSELs are less than 10.
- Manufacturability and economical yield.
- Scalability: For high-power applications, VCSEL's can be directly processed into monolithic 2D arrays.

Exercise 5.23

Which advantages does a VCSEL have?

▶ **Tip**
Help H5-23 (Sect. 12.1)
Solution S5-23 (Sect. 12.2)

Modulation of Laser Light

6

6.1   Task and Problems of Laser Modulation

In general, we have analog (continuously changing) and digital signals (a combination of 0 and 1 bits). Today most data are transformed to digital electrical signals of voltage V. For optical communication we have to transfer these digital electrical signals into optical ones. The analog signal (continuous in time and amplitude) must be digitized in time - this is done with the help of bits. Time is divided into individual sections, and a certain number of bits is placed per time section. This results in the bit rate as the number of bits per time unit (unit of measurement: bits per second, bps). In addition, the amplitude must be digitized, i.e., represented by bits. This means that each piece of information is represented by a combination of 0 or 1 bits and transmitted digitally [Brü 22].

These signals are now to be transmitted into the optical domain, in the simplest form as modulation of the laser power P (intensity modulation); a 0-bit then corresponds in the simplest case to the power P close to zero, while the 1-bit corresponds to a non-zero power (Fig. 6.1).

The bases for optical transfer are digital signals in the form of bits. Usually, digital signals are described as rectangular pulses, but this description is physically not correct due to points of discontinuity (marked in a). Indeed, ideally rectangular *electrical* pulses can be transformed into nearly rectangular *optical* pulses (bits), but during propagation in fibers of particular lengths, these pulses will be deformed to nearly Gaussian pulses (Fig. 6.2). The shorter the pulse length the shorter the fiber length which is necessary for this deformation. Depending on the bit transfer rate a single bit has a Gaussian shape

Supplementary Information The online version contains supplementary material available at https://doi.org/10.1007/978-3-658-43242-3_6.

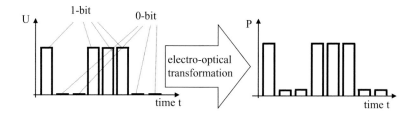

Fig. 6.1   Digital electrical and optical signals

Fig. 6.2   Ideal (a) bit and bits
at high transfer rates (b)

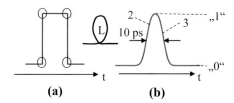

(a)   (b)

(Fig. 6.2b - at high transfer rates of about 10–100 Gbps). The "1" in Fig. 6.2 denotes the logical binary 1, and the "0" denotes the logical binary 0. The markers 2 and 3 in Fig. 6.2 represent the rising and falling edges, respectively.

6.2   Modulation Methods in Optical Communications Engineering

Amplitude and time can be described as continuous (analog) changes (where any value is possible) or as discontinuous (discrete or digital) changes (where only certain discrete values are permitted).

6.2.1   Amplitude Modulation, Power Modulation (AM, PM)

At amplitude modulation (AM), changes of the field strength of a carrier wave as well as changes in time are described continuously (Fig. 6.3). Mathematically, this is the modulation of the amplitude of a carrier frequency f_c with a modulation frequency f_m

$$A(t) = A_0 \cos 2\pi f_c t \cdot \cos 2\pi f_m t \qquad (6.1)$$

where A_0 is the maximum amplitude.
Using the product rule of two sine functions:

$$\sin \alpha \cdot \sin \beta = \frac{1}{2}[\cos (\alpha - \beta) - \cos (\alpha + \beta)]$$

Fig. 6.3 Amplitude and power modulation of a carrier wave (analog)

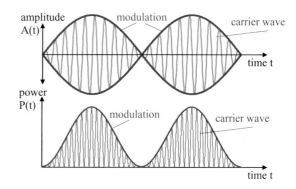

we get instead of the following equation:

$$A(t) = \frac{A_0}{2}[\cos 2\pi(f_c - f_m)t - \cos 2\pi(f_c + f_m)t]$$

From this representation of the amplitude modulation, it becomes clear that the modulation creates so-called sidebands $f_c - f_m$ and $f_c + f_m$ in addition to the carrier frequency f_c. This means that the bandwidth is increased.

In optical communication technology where *power* rather than amplitude plays the key role, we use the term *power modulation* (PM). Note that power or intensity is approximately connected with amplitude squared.

$$P(t) \propto A(t)^2 = A_0^2 \left(\cos 2\pi f_c t \cdot \cos 2\pi f_m t\right)^2 \tag{6.2}$$

Example of calculations in MathCad is given in the folder "Extras" (also exists as pdf file). Thus, the analog, temporal course of the amplitude is created according to Fig. 6.3.

6.2.2   Pulse-Amplitude Modulation (PAM)

At pulse-amplitude modulation, changes in time are described discretely (pulse-shaped, e.g., equidistant rectangular pulses), whereas changes of the field strength are described continuously as changes of the amplitude of pulses (Fig. 6.4).

Fig. 6.4 Pulse-amplitude modulation

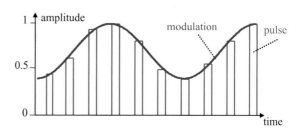

6.2.3   Pulse Position Modulation (PPM)

Pulse-position modulation is one kind of phase modulation. Changes of field strength are described discontinuously (i.e., fixed values for amplitudes, e.g., binary), whereas changes in time are described continuously (Fig. 6.5). In this case, the shift of the bit position (position modulation) describes the amplitude.

   Instead of a pulse position, one can also change the carrier frequency "clockwise" with modulation frequency (frequency modulation).

6.2.4   Pulse-Code Modulation (PCM)

Up to now we have described *analog* methods. The pulse-code modulation (PCM) is a purely digital method, where the field strength as well as changes in time are described discontinuously (i.e., stepwise). PCM (see also Sect. 2.4) is primarily a method of encryption. Only coded (mostly binary) signals are subject to modulation. The PCM method of encryption is sketched in Fig. 6.6.

   At certain points in time (i.e., discontinuously in time) the continuous signal (solid line in Fig. 6.6) is analyzed: the corresponding amplitude is in one of 8 positive or 8 negative ranges. The result will be described binarily, i.e., the corresponding code first contains the sign (+ is 1, − is 0) and then as the binary description of the range (e.g., 111 for the range 7, 010 for the range 2 or 000 for the range 0). Thus, for description of *one* amplitude value we need 4 bits (e.g., 1101 for the range +5). It results in higher requirements for the transfer bandwidth.

Fig. 6.5 Pulse-position modulation

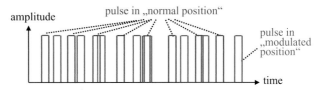

Fig. 6.6 Encryption of signals using the PCM method

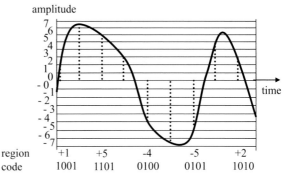

6.3   Direct Modulation of Semiconductor Lasers

At *direct modulation* the electric current in flux direction I_F is modulated. Because current I_F corresponds to a certain optical power, direct modulation is connected with the P-I characteristic (see Fig. 5.15). The modulating signal is mirrored at the laser part (above threshold current I_{th}) of this characteristic (Fig. 6.7).

Power of the modulated signal p_M depends on the selection of the working point by a corresponding initial current I_i with respect to the threshold value I_i/I_{th} together with the modulation depth $\Delta I/I_{th}$. Furthermore, lifetime of the carriers (electrons and holes) in semiconductor τ_L and of light particles (photons) τ_p is of importance. Power can be calculated by

$$p_M = 10 \cdot \log\left[\frac{\Delta I}{I_{th}} \cdot \frac{\frac{I_i}{I_{th}} - 1 - \omega^2 \tau_c \tau_p}{\left(\frac{I_i}{I_{th}} - 1\right)^2 - \omega^2 \tau_p \tau_c \cdot \left(\frac{I_i}{I_{th}} - 1\right) + \omega^2 \tau_c^2 \frac{I_i^2}{I_{th}^2}}\right] \quad (6.3)$$

where $M = \Delta I/I_{th}$ is the modulation depth, $V = I_i/I_{th}$ is the bias current, τ_c is the carrier lifetime and τ_p is the photo response lifetime.

For $\Delta I/I_{th} = 0.25$, $\tau_c = 10^{-9}$ s, and $\tau_p = 10^{-11}$ s one can calculate the dependence of the level of the modulated signal p_M on the (circular) frequency $\omega = 2\pi f$ (f is the frequency) for different initial currents I_i/I_{th} between 1.05 and 1.5. Results are shown in Fig. 6.8. Details of calculations are given in MathCad in the folder "Extras", where you can also find the corresponding pdf file.

One sees from (6.3) that there is a zero for the denominator of (6.3)

$$(V - 1)^2 - \omega_{max}^2 \tau_c \tau_p \cdot (V - 1) + \omega_{max}^2 \tau_p^2 V^2 = 0.$$

So, one considers "only" angular frequencies $\omega < \omega_{max}$

$$\omega_{max} = \frac{V - 1}{\sqrt{(V - 1) \cdot \tau_p \cdot \tau_c - \tau_p^2 \cdot V^2}} \quad (6.4)$$

Fig. 6.7 Modulation and P-I characteristic

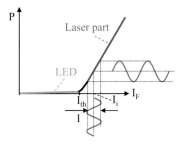

Fig. 6.8 Dependence of power p_M of modulated signals on frequency ω

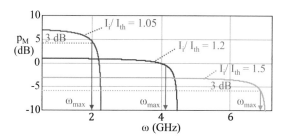

The transfer bandwidth is defined as the frequency of 3 dB power drop. That means, the higher the initial current the higher the transfer bandwidth for analog signals. From Fig. 6.8 together with some other calculations, one can deduce the following Table 6.1:

A practical example of a semiconductor laser SLD 202 V (Sony) is depicted in Fig. 6.9. At power 7 mW we get as maximum modulation frequency of 1.2 GHz, at 14 mw already 1.5 GHz. It means that transfer bandwidth can change with the laser power. This is important for practical use.

Exercise 6.1

Calculate the 3 dB drop in bandwidth for the example given in Fig. 6.9!

Table 6.1 Transfer bandwidth at different initial currents

Initial current I_i / I_{th}	3 dB transfer bandwidth (GHz)
1.05	2.0
1.1	3.0
1.2	4.3
1.3	5.3
1.4	6.2
1.5	6.9

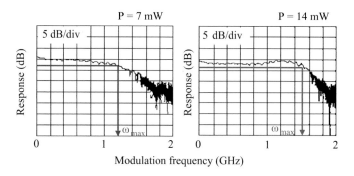

Fig. 6.9 Transfer bandwidth of a laser diode SLD 202 V

▶   **Tip**
Help H6-1 (Sect. 12.1)
Solution S6-1 (Sect. 12.2)

Other important effects at direct modulation result from the existence of a threshold current I_{th} in the laser characteristic.

At stepwise increase of current (Fig. 6.10a) one can see the following:

- The beginning of the bit-like current pulse of length τ_p and modulation depth ΔI result in the generation of electrons and holes (electron-hole pairs in Fig. 6.10b) as a precondition to reach the laser threshold. This part of the current pulse is "lost" for the emission of the light pulse; we get a delay time Δt between the beginnings of the current and light pulses (Fig. 6.10c).

$$\Delta t = \tau_P \cdot \ln\left(\frac{\Delta I - I_i}{\Delta I - I_{th}}\right) \tag{6.5}$$

At $I_i = I_{th}$ we get $\Delta t = 0$ - this is the optimum initial current.

- combination of electrons with holes, and laser light is emitted. Therefore, the number of electron-hole pairs is reduced. On the other hand it results in a reduced light intensity - maybe the laser is "switched off". Because we have more and more current resulting in new electron-hole pairs, we get "new" laser light. It results in so-called relaxation oscillations, which always occur at the beginning of the laser pulse (Fig. 6.10c).

In summary, the generation of light pulses is connected with a time delay Δt and relaxation oscillations.

Time delay as well as relaxation oscillations can be described in terms of so-called rate equations. These *coupled differential equations* describe both the time changes in electron-hole pair density N_{EHP} (increase of density by current I, decrease by induced emission of light) and changes in the light intensity P (increase by induced emission of

Fig. 6.10 Transient and relaxation oscillations at direct laser modulation

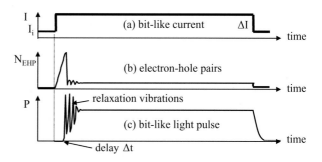

light, decrease of density by losses κ in the laser resonator). Thus, we get the following system of equations:

$$\frac{dP}{dt} = -\kappa \cdot P + a_0 \cdot N_{EHP} \cdot P$$

$$\frac{dN_{EHP}}{dt} = I - a_1 \cdot N_{EHP} \cdot P \tag{6.6}$$

a_0 and a_1 are constant values connected with the concrete laser construction. Such a set of differential equations can be solved numerically, e.g., by Mathcad (see also pdf file). One of these calculations for $I_i > I_{th}$ is depicted in Fig. 6.10.

Exercise 6.2

Which problems can result from the nonlinear characteristic of a laser? Do we also find these problems in an LED?

▶ **Tip**
   Help H6-2 (Sect. 12.1)
   Solution S6-2 (Sect. 12.2)

From rate equations, one can find the transfer function H (ω) for intensity, which contains resonance frequency ω_R and attenuation constant γ:

$$H(\omega) = \frac{\omega_R^2}{\sqrt{\left(\omega_R^2 - \omega^2\right)^2 + (\omega\gamma)^2}} \tag{6.7}$$

The transfer function H (ω) can be calculated using the parameter γ/ω_R. Results for selected values of γ/ω_R are shown in Fig. 6.11.

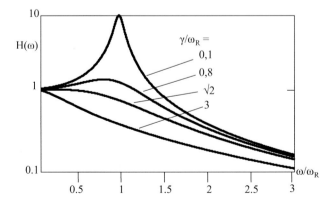

Fig. 6.11 Transfer function at intensity modulation

Fig. 6.12 Time delay Δt and relaxation oscillations (practical example after [Opi 95])

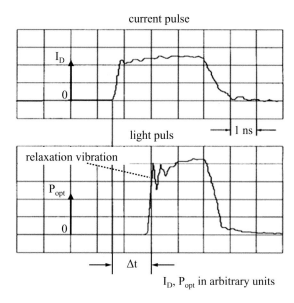

current pulse

I_D

0

light puls $\longmapsto$ 1 ns $\longleftarrow$

relaxation vibration

P_{opt}

0

$\longrightarrow$ Δt $\longleftarrow$

I_D, P_{opt} in arbitrary units

The results of this simulation can be proven experimentally (Fig. 6.12). At $I_i > I_{th}$ from a 4 ns current pulse (i.e., bit duration of about 4 ns) we get with time delay Δt a 2.5 ns light pulse. This pulse shortening must be considered in practice.

These results, obtained by simulation, can be well verified in experiments (Fig. 6.13).

For higher transfer bandwidth, one has to use shorter current pulses, e.g., for transfer rates above 1 Mbps one needs < 1 ns current pulses. For a sequence of short pulses, we have the same restrictions for time delay and relaxation oscillations. At the initial current *below threshold,* the current pulse is used completely to generate electron-hole pairs and we have *no* optical pulse. With a bias current above the threshold, the influence of the relaxation oscillations becomes apparent. This so-called bit-pattern effect is depicted in Fig. 6.13 [Opi 95].

Bit-pattern effect occurs especially if the initial current I_i is less than or equal to the threshold current I_{th}. At $I_i = 0.95\, I_{th}$ the first two bits "disappear", and the other bits are shortened (Fig. 6.13b). At $I_i = I_{th}$ the first two bits are shortened, whereas at the other bits, relaxation oscillations appear (Fig. 6.13c). At $I_i = 1.05\, I_{th}$ all light pulses show relaxation oscillations (Fig. 6.13d). To avoid bit-pattern effect the initial current should be chosen considerably higher than the threshold value.

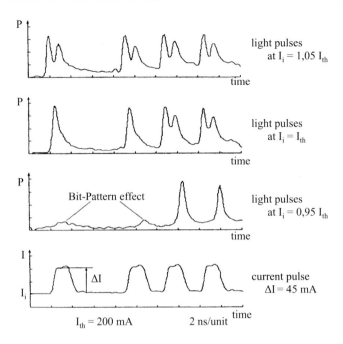

light pulses
at $I_i = 1{,}05\ I_{th}$

light pulses
at $I_i = I_{th}$

Bit-Pattern effect

light pulses
at $I_i = 0{,}95\ I_{th}$

ΔI

current pulse
$\Delta I = 45$ mA

$I_{th} = 200$ mA   2 ns/unit

Fig. 6.13 Bit-Pattern effect

6.4   External Modulation of Semiconductor Lasers

We can avoid problems in direct modulation such as time delay and relaxation oscillations using *external modulation*. In this case, a cw (continuous waves) laser is modulated (Fig. 6.14).

External modulation is based mainly on *linear electro-optical effect* (Pockels effect, details see Sect. 8.1.2).

6.4.1   Phase and Frequency Modulation

For phase and frequency modulation, the linear electro-optical effect is mainly used. The simplest arrangement of a phase modulator consists of a wave guiding strip line. This strip line is fabricated by corresponding doping of an electro-optical substrate (mostly

Fig. 6.14 Basics of external
modulation

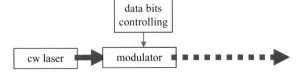

data bits
controlling

cw laser   modulator

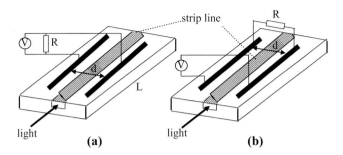

Fig. 6.15 Phase modulator with voltage applied in the middle (a) and traveling wave phase modulator (b)

LiNbO$_3$, Fig. 6.15). Applying a voltage V to both sides of the strip line, we get changes of the effective refractive index Δn_{eff} and therefore of the phase $\Delta \varphi$.

   If light wave propagates along the length L, we get a phase rotation $\Delta \varphi$:

$$\Delta \varphi = \frac{2\pi}{\lambda} \cdot n_{eff} \, L \tag{6.8}$$

With a phase modulator one can realize, e.g., a pulse-phase modulation.

   Maximum modulation frequency is determined by the kind of voltage supply. At phase modulator with voltage applied in the middle (Fig. 6.15a) the maximum modulation frequency of about some GHz is determined by the time constant given by capacity C of electrodes and terminating resistor R (mostly 50Ω). At traveling wave phase modulator (Fig. 6.15b), the electric field propagates together with the light. Electrodes act as a line with terminating resistor R. No reflection of the electric field takes place and one can achieve higher modulation frequencies (theoretically more than 1 THz; practically about 40 GHz).

   For frequency modulation, one has to use two-phase modulators in an interferometric arrangement (e.g., as Mach-Zehnder interferometer as depicted in Fig. 6.15).

6.4.2   Power Modulation

Intensity modulation or power modulation of light can be realized in different ways - one can modulate emission, absorption, or bifurcation of light.

   At emission modulation mode, propagation is constricted by an electric field. A basic mode just propagates in the strip line *without* electric field. *With* electric field we get a reduced effective refraction index. Due to the worsening of light propagation in the strip line, light will be radiated. In practice this method is hard to realize, which is why it has not been used up to now.

At electro-absorption one uses the shift of the absorption edge in semiconductors by an electric field (Franz-Keldysh effect). Transparent semiconductors without voltage become absorptive with an electric field. In practice, early tests are being performed using thin semiconductor layers of a few nanometer thickness (so-called quantum films). Bit rates of more than 20 Gbps have been achieved.

Often Mach-Zehnder interferometers (see Fig. 8.4 in Sect. 8.1.2) are used. With 5 V one can modulate in crystalline $LiNbO_3$ with about 75 GHz.

Another possibility for external intensity modulation is the use of *controllable symmetrical directional couplers*. In a passive coupler (Fig. 6.16a) of proper dimensions (distance s and width d of waveguides, coupling length L), light will be bifurcated to different gates. At the beginning light wave is completely in waveguide 1; after the distance L/4 light is split into 2 equal portions in waveguide 1 and 2; after L/2 light is completely in waveguide 2; after 3L/4 again 2 equal portions in waveguides 1 and 2; after distance L light is again completely in waveguide 1. Depending on the distance L we can get light either at gate 3 or 4, or light can be distributed between gates 3 and 4 in a certain relation (Fig. 6.16b). To realize a passive coupler, we should keep certain definite relations between s and L.

In a *controllable symmetrical directional coupler* without an electric field, light propagates without any losses from gate 1 to gate 3; with an electric field, the light is completely coupled to gate 4 (Fig. 6.17). Thus, at gate 4 we get the modulated signal, whereas at gate 3 we have the inverted signal.

Fig. 6.16 Passive symmetric directional coupler

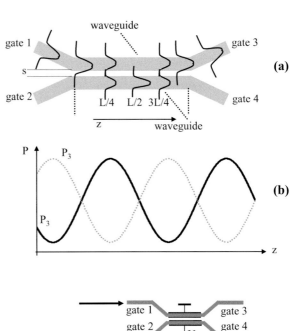

Fig. 6.17 Directional coupler as modulator

Exercise 6.3

How does modulation by symmetrical directional coupler work?

▶   **Tip**
Help H6-3 (Sect. 12.1)
Solution S6-3 (Sect. 12.2)

Optical Receivers

7

Optical receivers have to transfer light into *electrical quantities* in order to be measured. These electrical data can be displayed or processed. Therefore, receivers have the following requirements:

- *High sensitivity*—while high sensitivity is usually desired, it can result in errors (bit-error rate). The lower the requirements with regard to bit errors, the higher is the sensitivity of optical receivers.
- *High speed*—a narrow (optical) bit pulse must be transferred to an electrical pulse as fast as possible. The speed is connected to the transfer bandwidth.
- *Low noise*—to achieve a low bit rate error.

7.1 Receiver Principles

For receivers in optical communication the *inner photo-electrical effect* is used. This means that the light of an adapted wavelength λ or frequency f (with $hf \geq E_g$, where E_g is again the band gap) brings electrons from the valence band (v-band) to the conductivity band (c-band, Fig. 7.1). These carriers then move into the direction of the voltage applied (electrons to the plus direction, holes to the minus direction). These carriers can be "counted".

In view of optical communication, we will describe those receivers that are currently most important—the pin and avalanche photo diodes receivers. Last but not least we will have a look at heterodyne receivers.

Supplementary Information The online version contains supplementary material available at https://doi.org/10.1007/978-3-658-43242-3_7.

© Springer Fachmedien Wiesbaden GmbH, part of Springer Nature 2024
V. Brückner, *Elements of Optical Networking*,
https://doi.org/10.1007/978-3-658-43242-3_7

Fig. 7.1 Inner photo-electric
effect in semiconductors

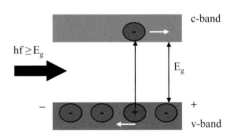

The inner photo effect appears in all semiconductors, but the spectral range is different for each semiconductor: the photo effect is related to wavelength λ of incoming light (in μm) and energy gap E_g (in eV):

$$\lambda^{\mu m} \leq \frac{1,24}{E_g^{eV}} \tag{7.1}$$

For example, Si ($E_g = 1.12$ eV) receiver can only be used for the first optical window, Ge ($E_g = 0.66$ eV), and InGaAsP ($E_g = 0.7 - 1.4$ eV) for the second and third optical window.

Simple semiconductor photoconductors are not suitable for optical communication - they are too slow and not sensitive enough. For this reason, photoconductors with a p-n-junction and a reversed-biased voltage (Fig. 7.2) are used where we get a depletion layer suitable for light absorption.

In the depletion region (about 20 μm thickness) we get a light-induced density of electrons and holes moving away from p-n-junction; this results in a photovoltage over the load resistance R_L. For the speed of motion of electrons (and holes) we have two possibilities:

1. The very slow diffusion with the diffusion speed v_{Diff}; diffusion is defined by the gradient of carrier concentration N and results in a reduction of gradient as depicted in Fig. 7.3 for electrons. Mathematically using a differential equation, one can describe the diffusion process as changes in the electron density N in time given by gradient $\frac{d^2N}{dx^2}$ and diffusion constant D:

$$\frac{dN}{dt} = D \cdot \frac{d^2N}{dx^2} \tag{7.2}$$

Fig. 7.2 p-n junction as
photoconductor and field
strength distribution

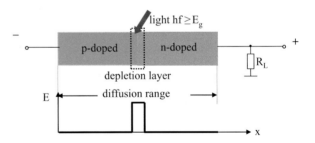

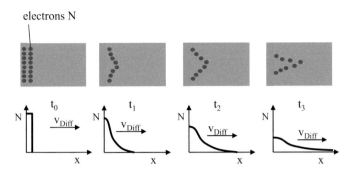

Fig. 7.3 Diffusion process in a pn diode

The diffusion constant D is, e.g., in germanium $150 \, cm^2/s$ and in silicon $20 \, cm^2/s$. In Fig. 7.3 the spatial distribution of electrons (gradient) is depicted at different times t_0 to t_3. This distribution shows a tendency to a balance. Thus, the speed of electrons declines.

2. Much Faster is the *Drift*, where electrons move in an electric field. The drift speed v_{Drift} of electrons depends on the field strength E:

$$v_{Drift} = \mu \cdot E \qquad (7.3)$$

The carrier mobility is μ. In general, the electron mobility is much higher than the hole mobility. For example, in Si the electron mobility is $\mu_e = 1{,}350 \, cm^2/Vs$, for holes only $\mu_h = 480 \, cm^2/Vs$, in GaAs $\mu_e = 8{,}500 \, cm^2/Vs$ and $\mu_h = 435 \, cm^2/Vs$, in Ge $\mu_e = 3{,}900 \, cm^2/Vs$ and $\mu_h = 1{,}900 \, cm^2/Vs$. Field strength is the voltage decay over the length. Because we have the voltage decrease along the depletion region (i.e., along $20 \, \mu m$) we have a drift speed in this region, where $v_{Drift} \gg v_{Diff}$. That means that the drift dominates in the depletion region, and diffusion dominates outside. This is why p-n-photo diodes are slow and not suitable to transfer high transfer rates.

7.2   pin Diode

The basic idea of a pin diode consists in a "prolongation" of the drift region by insertion of a non-doped (or intrinsic) semiconductor layer (i-layer) between p- and n-region. A potential drop takes place between p- and n-doped layers; thus, we have high field strength in this region and drift dominates compared with diffusion. Because drift range in a pin diode is much larger than the diffusion range (about $100 \, \mu m$), drift velocity dominates in pin diodes, and they become faster.

Pin diodes work at about 5 V; thus, one can use a power supply from a PC (TTL level). If we take $100 \, \mu m$ as the typical length of the intrinsic range (i-range) we get a field strength of 500 V/cm. Pin diodes are the fastest commercially available photo

Fig. 7.4 Design of a pin diode
and field strength distribution

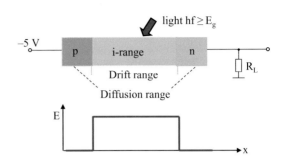

receivers. They show response times down to the picosecond range, and therefore they can be used for extremely high transfer rates.

Another advantage of pin diodes is the high linearity and therefore in a high dynamic range of about 10^7. That means that a 10^7 higher optical signal results in a 10^7 higher electrical signal. Of course, this is true only for weak signals.

Besides irradiation directly to the i-range like depicted in Fig. 7.4 one can also use an incidence of light through the narrow (some nanometer) p-range.

Summarizing one can say that pin diodes have a number of advantages:

- high speed, i.e., highest bit rates
- sufficiently high sensitivity
- high dynamic range
- low voltage (TTL level)
- low noise.

7.3   Avalanche Photo Diode (APD)

In an **a**valanche **p**hoto **d**iode (APD) we find a combination of high speed by drift and very high sensitivity. The design of an APD is depicted in Fig. 7.5a. In Si-APDs we get a potential drop along the pn junction up to $U = 300$ V. If one takes the thickness of the pn junction as $b = 10$ µm one gets a field strength $E = U/b = 300$ kV/cm, which is close to the field strength of breakthrough (destruction field strength). This value is, e.g., for GaAs about 320 kV/cm. Field strength distribution in an APD is shown in Fig. 7.5b. High field strength in the range of pn junction is followed by low field strength in i-range, resulting in a fast transport of electrons by drift.

Avalanche effect is illustrated in Fig. 7.5c. Light generates electron-hole pairs. Each single electron is accelerated in an electrical potential. Theoretically electron can get energy of motion (kinetic energy) of 300 eV—that means a multiple of gap energy (less than 1 eV). This accelerated electron is now able to create another electron-hole pair, and its kinetic energy will be shaved off about 1 eV. The newly created electron will be

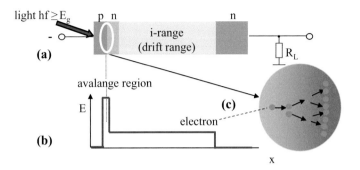

Fig. 7.5 APD design (a), field strength distribution (b), and avalanche process (c)

accelerated again resulting in another electron-hole pair, etc. This avalanche process runs until the electron reaches the barrier of the p-n range. Avalanche process results in the generation of many electrons starting from a single one—this explains the very high sensitivity of APDs.

Therefore, in an APD the primary photo current is amplified by a multiplication factor M.

$$M = \frac{1}{1 - \left(1 - \frac{U}{U_{Br}}\right)^{\kappa}} \tag{7.4}$$

with V—potential drop at pn junction, V_{Br}—breakthrough voltage and κ—material specific value (between 1.5 and 8). For a high amplification U must be close to destruction voltage V_{Br}. In germanium one uses about 40 V, in silicon between 150 and 200 V. Multiplication factors of 1000 can be reached, e.g., in Si-APDs (Fig. 7.6).

Electron-hole pairs are created also in other parts of the APD, e.g., in i-range. Holes move in the direction of the pn junction, will be accelerated there and will start another avalanche process. Due to the longer run time of holes (and also the extra time needed to go from i-range to the pn junction), the avalanche starts with a time delay, and we get a "smearing" in time and a prolongation of the electrical pulse - that means the speed of an APD is reduced compared to pin diodes.

Fig. 7.6 Dependence of multiplication factor on block voltage

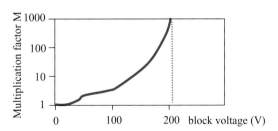

For APDs we can conclude:

Advantages:

- Highest sensitivity.
- Sufficiently high speed.

Disadvantages:

- High voltage necessary (up to 300 V in Si-APDs), therefore an independent power supply is necessary.
- Extra noise.
- Lower dynamic range.

Exercise 7.1

A receiver is able to detect -25 dBm. Express this value in mW!

▶ **Tip**
Help H7-1 (Sect. 12.1)
Solution S7-1 (Sect. 12.2)

7.4   Noise in Receivers, Bit-Error Rate (BER)

Photo current I_0 detected by the receiver is in fact an averaged value resulting from a fluctuating photo current $I(t)$. These fluctuations are demonstrated in Fig. 7.7a. For illustration we took a randomized "digital" noise, i.e., values of photo current change by multiples in whole numbers of "one" (in our example between 6 and 14). Then we get the averaged photo current as the averaged value of $I(t)$:

$$I_0 = \overline{I(t)} \tag{7.5}$$

Deviations from average value $I(t)$—I_0 over the time are depicted in Fig. 7.7b. That means, the "starting" current in a receiver can be treated as superposition of a DC with a noise current. The resulting calculation of the square of averaged photo current is shown in Fig. 7.7c. With values of Fig. 7.7a one gets as square of averaged photo current the numerical value 5.8 (dotted line in Fig. 7.7c):

$$\overline{i_R^2(t)} = \overline{(I(t) - I_0)^2} \tag{7.6}$$

For so-called *white noise*, i.e., for homogeneous distribution of noise over all frequencies we get the noise power density related to a certain bandwidth df of the photo receiver

$$\frac{\overline{di_R^2}}{df} = 2eI_{R0} \tag{7.7}$$

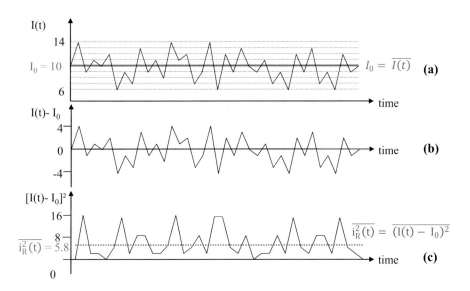

Fig. 7.7 Fluctuations of photo current (a and b) and averaged square of noise current (c)

with I_{R0}—averaged noise current and $e = 1.6 \cdot 10^{-19}$ As (elementary charge).

Thus, the square of the averaged noise current for white noise is

$$i_R^2 = 2eI_{R0} \, B \qquad (7.8)$$

with B—bandwidth (in good approximation one can take the 3 dB bandwidth of photo receivers).

Exercise 7.2

What means "noise is statistically"? What means "white noise"?

▶ **Tip**
Help H7-2 (Sect. 12.1)
Solution S7-2 (Sect. 12.2)

Now we want to explain the most important mechanisms of noise more detailed.

7.4.1 Shot Noise

Shot noise we get due to direct current in a p-n junction poled in reversed bias. Reasons for this DC at shot noise can be different:

- Due to quantum nature of light photon flux is not continuous, furthermore the positions of generation of electron-hole pairs in receivers are distributed statistically (quantum noise i_Q).
- By light generated electrons in the c-band can go back to the valence band due to random recombination processes (or holes in the v-band can go to c-band) and are therefore "loosed" (background recombination noise i_B).
- Even without light electron-hole pairs can be generated (dark current noise i_D).

In shot noise all these noise processes are combined to a shot noise current i_{shot}:

$$i_{shot} = \sqrt{i_{shot}^2} = \sqrt{i_Q^2 + i_B^2 + i_D^2} \qquad (7.9)$$

Exercise 7.3

What is the reason for shot noise?

▶ **Tip**
Solution S7-3 (Sect. 12.2)

7.4.2 Intensity Noise

Beside quantum noise laser light (combination of spontaneous and stimulated emission) has statistical nature - the intensity noise. At spontaneous emission of LED intensity noise can be neglected compared with other noise processes. In a semiconductor laser starting with spontaneous emission intensity fluctuations will be amplified. Furthermore, we have a "competition" between different longitudinal modes - the intensity distribution between different modes is changing permanently. Mathematical description of intensity noise is very complicated; there are no "manageable" formulas. Thus, intensity noise is mostly neglected.

7.4.3 Thermal noise, Nyquist noise

In any real resistance we find thermally induced motion of electrons Thus due to inhomogeneous distribution of electrons we find a statistical fluctuation between space charge regions resulting in a potential difference. That's why we find - even without external resistance - a permanently changing effective voltage $U_{th,\ eff}$ - the resistance "sweeps" thermally due to thermal or Nyquist noise:

$$U_{th,eff}^2 = 4k_B T \cdot R \cdot B \qquad (7.10)$$

where $k_B = 1.380662 \cdot 10^{-23}$ J/K is the Boltzmann constant, R is the real "noise" resistance, and B is the bandwidth of receiver. Thus, one can imagine the real noise resistance (Fig. 7.8a) as serial connection of an ideal (noise-free) resistance R and a noise source of voltage $U_{th,eff}$ (Fig. 7.8b). This serial connection of voltage sources can be described also by the parallel connection of current sources (Fig. 7.8c).

The ideal resistance R is given by path resistance R_B and load resistance R_L ($R = R_L + R_B$). Thus, thermal noise current I_{th} is given by

$$I_{th}^2 = i_{th,eff}^2 = 4k_B T \frac{B}{R_L + R_B} \tag{7.11}$$

The higher $R = R_L + R_B$ and/or the lower the temperature the lower is the thermal noise. That's why for highest requirements the receiver should be cooled down to temperature of liquid nitrogen (77 K). Even in amplifiers with this noise temperature we find thermal noise.

Exercise 7.4

What is the reason for thermal noise? How can we reduce thermal noise?

▶   **Tip**
Help H7-4 (Sect. 12.1)
Solution S7-4 (Sect. 12.2)

Total noise current i_{ges} in non-amplifying photo receivers (e.g., in pin diodes) is given by

$$i_{ges}^2 = i_{th}^2 + i_{shot}^2 = i_{th}^2 + i_Q^2 + i_B^2 + i_D^2 \tag{7.12}$$

All parts of noise result in a **n**oise **e**quivalent **p**ower (NEP) which is mainly determined by the load resistance R_L and receiver bandwidth B. Signal power P_S (in mW) related to NEP (also in mW) results in the **s**ignal-**n**oise **r**atio (SNR), measured in decibel (dB):

$$SNR^{dB} = 10 \cdot \log \frac{P_S}{NEP} \tag{7.13}$$

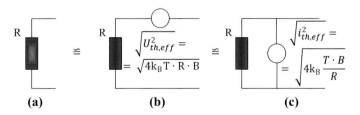

Fig. 7.8   Noise equivalent network

Fig. 7.9 Simplified high-frequency equivalent network for a receiver

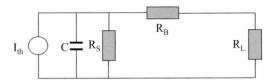

Finally, we get a simplified high-frequency equivalent network of a receiver (Fig. 7.9) with a capacity of barrier layer C and a resistance of barrier layer R_S.

In general $R_S \gg R_L$ is true; thus, we get the threshold frequency of receiver f_g as

$$f_g = \frac{1}{2\pi \cdot C \cdot (R_L + R_B)} \tag{7.14}$$

Typical receivers have a threshold frequency of 1–3 GHz, best receivers work up to 100 GHz.

7.4.4   Multiplication Noise

The current in APDs is amplified by the multiplication factor M, thus shot noise is also amplified and we get an extra noise $i_{zusatz, APD}$ in APDs increasing with the amplification as

$$i^2_{zusatz,APD} = i^2_{shot} \cdot M^2 \cdot F_D \tag{7.15}$$

The extra noise factor $F_D = M^x$ reflects the statistical (stochastic) character of carrier multiplication in an APD. The value of x depends on optical frequency, receiver material, and multiplication factor M, for silicon it is $x = 0.2 - 0.5$, for germanium $x = 0.8 - 1$. For mixed crystals InGaAsP we get $x = 0.6 - 0.8$. Thus, we get the amplified shot noise in an APD:

$$i^2_{shot,APD} = i^2_{shot} \cdot M^{2+x} \tag{7.16}$$

and the total noise in an APD:

$$i^2_{ges} = i^2_{zusatz,APD} + i^2_{th} \tag{7.17}$$

The signal-noise ration SNR in an APD is *downgraded* by extra noise.

All noises (shot noise, extra noise in APD, and thermal noise in receivers and amplifiers) are summarized in an equivalent network (Fig. 7.10).

Exercise 7.5

What kind of noise we can expect in an APD receiver?

▶ **Tip**
   Help H7-5 (Sect. 12.1)
   Solution S7-5 (Sect. 12.2)

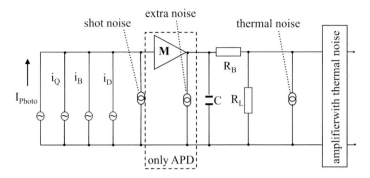

Fig. 7.10 Equivalent network with all noise sources

7.4.5   Bit-Error Ratio

Bit-error rate or more precisely **B**it-**E**rror **R**atio (BER) is the number of incorrect trans-ferred bits per time unit or per total number of bits transferred, respectively. In networks of Deutsche Telekom, a BER of 10^{-10} is required, that means, from 10^{10} bits on average 1-bit is allowed to be transferred incorrect. Of course, with increasing SNR the BER is decreased. This can be described approximately by formula

$$\text{BER} = \frac{1}{2} \cdot \text{erfc}\left(\sqrt{\frac{1}{8}\frac{P_S}{\text{NEP}}}\right) = \frac{1}{2} \cdot \text{erfc}\left(\sqrt{\frac{1}{8} \cdot 10^{\frac{\text{SNR}_{\text{dB}}}{10}}}\right) \qquad (7.18)$$

where erfc (x) is the so-called complementary error function:

$$\text{erfc}(x) = \frac{2}{\sqrt{\pi}} \cdot \int_{x}^{\infty} e^{-t^2}\, dt$$

In Fig. 7.11 the connection between bit-error rate BER and signal-noise ratio SNR is shown (Mathcad calculation, see also pdf file).

For a BER of 10^{-10} we get SNR $= 22$ dB and, therefore the proportion $P_S/$NEP $= 158$ - if we know NEP the necessary signal power can be calculated.

Exercise 7.6

How NEP, SNR, and BER are connected?

▶ **Tip**
   Solution S7-6

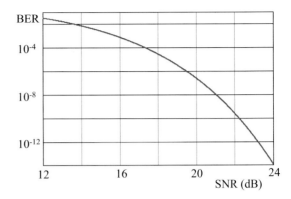

Fig. 7.11 Relation between BER and SNR

7.4.6 Optical heterodyne

A significant increase in the sensitivity of a receiver can be achieved using principles of superposition in a nonlinear element. This principle is well-known for a long time in radio engineering—super heterodyne receivers more or less replaced the forward receivers of former time. Basic idea is shown in Fig. 7.12.

Let's consider the interference of modulated transmitter light of frequency f_T (or of wavelength λ_T) with a local laser (at the place of receiver) of frequency f_L. At receiver we get a total time-dependent field strength E_{total} (t)

$$E_{ges}(t) = E_S(t) \cdot \cos\left(2\pi f_S\, t + \varphi_S\right) + E_L(t) \cdot \cos\left(2\pi f_L\, t + \varphi_L\right) \quad (7.19)$$

with $E_T(t)$ as amplitude modulated bit shaped, E_L as amplitude of local cw laser, and ϕ_T or ϕ_L as phase. A linear addition of field strengths cannot result in new frequencies at the exit of receiver other than f_T and f_L. Thus, one needs a nonlinear element.

In a receiver like pin diode or APD one can measure power as averaged value of field strength squared. With $E(t) = E_0 \cos(2\pi f_0 t)$ we get the current i_S as well as the power P as square of field strength averaged in time:

$$i_S \propto P \propto E^2(t) = \frac{1}{T} \cdot \int_t^{t+T} \overline{E^2(t)} dt = \frac{1}{T} \cdot \int_t^{t+T} \frac{E_0^2}{2} \left[1 + \cos(2\pi f_0 t)\right] dt = \frac{E_0^2}{2} \quad (7.20)$$

The duration of integration time T should be large compared with the period of optical signal. Then we have to integrate over total field strength $E_{total}(t)$ instead of $E(t)$.

Fig. 7.12 Basic idea of an optical heterodyne

$$i_S \propto \overline{E_{ges}^2(t)} = \overline{E_S^2 cos^2(2\pi f_S t + \varphi_S)} + \overline{E_L^2 cos^2(2\pi f_L t + \varphi_L)}$$
$$+ \overline{2E_S E_L cos(2\pi f_S t + \varphi_S)cos(2\pi f_L t + \varphi_L)}$$

(7.21)

Averaging by integration results in field strengths at the following frequencies:

frequency	power
$0 = f_L - f_L$	$\frac{E_L^2}{2}$
$0 = f_T - f_T$	$\frac{E_L^2(t)}{2}$
$2 f_L = f_L + f_L$	$\frac{E_L^2}{2} \cdot cos(4\pi f_L t + 2\varphi_L)$
$2f_T = f_T + f_T$	$\frac{E_T^2}{2} \cdot cos(4\pi f_T t + 2\varphi_T)$
$f_L + f_T :$	$\frac{E_T \cdot E_L}{2} \cdot cos(2\pi f_L + f_T)t + \varphi_L + \varphi_T$
$f_L - f_T$	$\frac{E_T \cdot E_L}{2} \cdot cos(2\pi f_L + f_T)t + \varphi_L + \varphi_T$

Receiver acts as low-pass filter. That means terms with the frequencies $2f_L$, $2f_T$, and $f_L + f_T$ are omitted, and we get only terms at frequencies 0 and $f_L - f_T$. Thus, we get the current i_S as

$$i_S \propto P_S(t) + P_L + 2\sqrt{P_S(t)P_L}cos\left[2\pi(f_L - f_S)t + \Delta\varphi\right]$$

(7.22)

Frequency $f_{IF} = f_L - f_T$ is the so-called intermediate frequency and $\Delta\varphi = \varphi_L - \varphi_T$ is the phase difference. Both f_L and f_T are in the range of 10^{14} Hz, its difference is typically less than 1 GHz. Different modulation methods (AM, FM, or PM) of transmitted signal result in different modulations of $\sqrt{P_T(t)}$, intermediate frequency f_{IF} or phase $\Delta\phi$.

Case $f_L \neq f_T$ or $f_{IF} > 0$ is named *heterodyne reception*, case $f_{IF} = 0$ *homodyne reception*, respectively.

In general $P_L \gg P_T$ is true because the light of local laser is transferred nearly loss-less to the receiver. That's why it is not recommended to use an APD. The gain which can be reached by heterodyne reception compared with direct reception by an APD (with optimized multiplication factor M_{opt}) depends on powers of signal P_S and of local laser P_L at receiver's entrance and can be up to 17 dB or more. For very weak signals APD is of advantage, for not too weak signals the heterodyne receiver.

In Fig. 7.13 an integrated-optical heterodyne receiver developed at Heinrich-Hertz-Institute Berlin is shown [Hei 94].

Note that one has to take care of the extremely strong dependence of heterodyne reception on the polarization of light. After transmission in a glass fiber polarization of signal light is not predictable and not plannable. Therefore, both signal and local laser light (in Fig. 7.13 a DBR laser) are split into TM- and TE-waves by a polarization split-ter. TM- and TE-waves will be coupled separately (interference part) and detected by separate pin diodes.

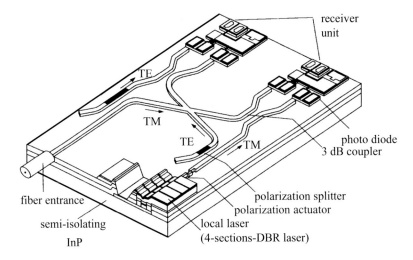

Fig. 7.13 Integrated-optical heterodyne receiver [Hei 94]

Using heterodyne reception best results one can get for phase modulation - it is possible to detect a single bit with about 20 photons. This is close to the theoretical limit of about 10 photons.

One advantage of heterodyne reception is the very high selectivity due to (commercially available) electric filters at intermediate frequency. Therefore, heterodyne receivers seem to be useful for DWDM signals. On the other hand, the use of (commercially available) optical pre-amplifiers (e.g., just in front of the receiver) nearly compensates for the sensitivity gain in heterodyne receivers.

One problem of heterodyne receivers compared with direct reception by pin diodes or APDs is the very high requirements on frequency stability of signal and local laser light at high bit rates (e.g., at DWDM operation). At amplitude and frequency modulation the laser frequencies instabilities less than 10% of the bit rate are required, at phase modulation even less than 0.01%.

Exercise 7.7

Which sensitivity of receiver (in Ws) we have to expect, if only the minimum necessary 21 photons in third optical window are available?

▶ **Tip**
Help H7-7 (Sect. 12.1)
Solution S7-7 (Sect. 12.2)

Compounds of Optical Networks

8

A network consists of many individual elements that can be combined into assemblies. Some of them (couplers and splitters) we already got to know.

An essential part of optical networks is filters and switches. For example, couplers and switches can be combined in such a way that different wavelengths are combined (multiplexing, MUX, or wavelength division multiplexing, WDM) or separated (demultiplexing, DEMUX), that wavelengths are decoupled or added (optical add-drop multiplexer, OADM) or that wavelengths are "recoupled" from one fiber to another (optical cross-connector, OCC). The filters and switches required for this are shown below.

Central to the transmission of digital signals in optical networks is the regeneration of bits in amplitude, shape, and clock - the so-called 3R technology. This leads until now to the treatment of optical amplifiers and methods of dispersion compensation.

8.1   Switches in Optical Networks

Switches are "controllable" couplers. In Fig. 8.1a light from gate 1 is switched to either gate 2 or gate 3 - this is the 1×2 switch (pronounced as: one on two switch). In the 2×2 switch from Fig. 8.1b either gate 1 is switched to gate 4 and gate 2 to gate 3 (cross-switching) or gate 1 is switched to gate 3 and gate 2 to gate 4 (bar-switching) - this is the crossbar switch. Figure 8.1c shows a $1 \times N$ star switch where the light from gate 1 can be switched to one of N gates. Decisive factors are the insertion loss, the crosstalk, and the switching time, which should not exceed 10 ms because of the associated optical or electrical buffering ("lost" bits).

Supplementary Information The online version contains supplementary material available at https://doi.org/10.1007/978-3-658-43242-3_8.

© Springer Fachmedien Wiesbaden GmbH, part of Springer Nature 2024
V. Brückner, *Elements of Optical Networking*,
https://doi.org/10.1007/978-3-658-43242-3_8

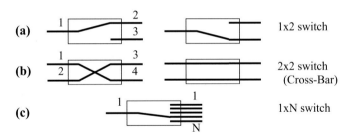

Fig. 8.1 Basic design of optical switches

Switching can be realized either mechanically, electro-optically, mechano-optically, or thermally.

8.1.1 Mechanical Switches

In a mechanical fiber switch, light from fiber 1 is switched to fiber 2 *or* 3 mechanically (Fig. 8.2).

The distance between fiber 1 and fiber 2 or 3 is very small (μm range). Therefore, light from fiber 1 can be coupled to fiber 2 or 3 with high efficiency—that means low insertion losses (0.8 to 3 dB) and crosstalk ($a_{ct} > 60$ dB) are very small. Otherwise, the switching time is more than a few milliseconds (sometimes up to 1 s). This limits its practical use.

8.1.2 Electro-Optical Switches

Electro-optical switching is based on the linear electro-optical effect, or Pockels effect [Ped 08]. Linearly polarized light impacts on a transparent nonlinear crystal (often $LiNbO_3$). If we apply a voltage V corresponding to a field strength E due to the nonlinear coefficient r, we get a rotation of polarization plane by an angle $\Delta\varphi$ (Fig. 8.3):

$$\Delta\varphi = \pi \cdot n^3 \cdot \frac{r}{\lambda} \cdot L \cdot E \tag{8.1}$$

n is the (linear, it means constant) refractive index, L is the length, and d is the thickness of the crystal, respectively.

Fig. 8.2 Mechanical fiber switch

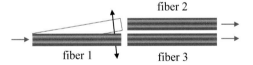

Fig. 8.3 Longitudinal (a) and transversal (b) Pockels effect

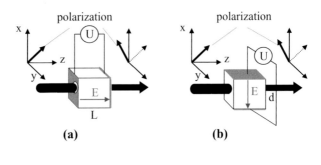

(a)   (b)

Fig. 8.4 Mach-Zehnder interferometer

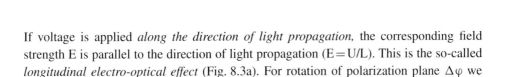

If voltage is applied *along the direction of light propagation,* the corresponding field strength E is parallel to the direction of light propagation (E=U/L). This is the so-called *longitudinal electro-optical effect* (Fig. 8.3a). For rotation of polarization plane $\Delta\varphi$ we get

$$\Delta\varphi = \pi \cdot n^3 \cdot \frac{r}{\lambda} \cdot U \tag{8.2}$$

If voltage is applied orthogonally to the direction of light propagation, the corresponding field strength E is also orthogonal to the direction of light propagation (E=U/d with d as the distance of electrodes). This is the so-called *transversal electro-optical effect* (Fig. 8.3b). In this case for rotation of polarization plane $\Delta\varphi$ we get

$$\Delta\varphi = \pi \cdot n^3 \cdot \frac{r}{\lambda} \cdot L \cdot \frac{U}{d} \tag{8.3}$$

For $LiNbO_3$ as a typical material the nonlinear coefficient r at wavelength 633 nm is about $36 \cdot 10^{-12}$ m/V, refractive index n $\cong$ 2,2 and typical crystal length L = 2 cm. For a rotation of polarization plane of $\Delta\varphi = \pi/2 = 90°$ one needs a voltage in the kV region for the longitudinal Pockels effect. In case of transversal Pockels effect, we need field strength in the range of some kV/cm, which can easily be realized by the voltage of a few volts along the crystal thickness of about 10 μm. Instead of the crystal thickness, we can select the distance between electrodes d to be very small.

Exercise 8.1

Which voltage is needed to rotate the polarization by $\Delta\varphi = \pi/2$ by longitudinal Pockels effect if length of the $LiNbO_3$ crystal is 2 cm and wavelength is $\lambda = 1500$ nm?

▶ **Tip**
   Help H8-1 (Sect. 12.1).
   Solution S8-1 (Sect. 12.2).

Exercise 8.2

Which field strength is necessary if we use the transversal Pockels effect under the same conditions as in Exercise 8.1?

▶ **Tip**
   Help H8-2 (Sect. 12.1).
   Solution S8-2 (Sect. 12.1).

A typical application of Pockels effect is the Mach-Zehnder interferometer (

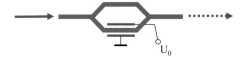

), where polarized light is first split by a Y-splitter into two parts of equal power. In a second Y coupler, this will be recombined (switch position "on"). In a perfect interferometer both wings have the same length. If we apply voltage, the phase of light in the wings will be changed and the phase propagation speed is different. In case of phase opposition (phase shift of λ/2), we can get an extinction of light (switch position "off"). A Mach-Zehnder interferometer (MZI) based on LiNbO$_3$ can be realized with 5 V switching voltage. Up to 75 GHz switching frequency can be used, i.e., it is very fast. It is important to note that the switching behavior (on-off ratio) very critically depends on the voltage applied and the polarization of light; therefore, high standards of configuration are required in the production process.

In combination with a frequency-selective directional coupler, one can use MZI as a 1×2 switch with controllable demultiplexing function (Fig. 8.5).

If one combines directional couplers (DC) and Mach-Zehnder interferometers (MZI) as shown in Fig. 8.6, a 2×2 crossbar switch is realized. Without voltages V_1 and V_2 the signal will be transferred from A to C and from B to D (bar); with voltage U_1 and U_2 from A to D and from B to C (cross). The advantage of these switches is the short switching time (ns range); the disadvantage is the relatively high insertion losses caused primarily by coupling light from glass fiber into the LiNbO$_3$ waveguide and vice versa.

Many 2×2 crossbar switches can be combined to create a switching matrix. For example, in Fig. 8.7, an 8×8 switching matrix is depicted. Of course, this kind of switching matrix is not scalable at will because, in addition to insertion losses, the relatively high crosstalk of 20–30 dB is a disadvantage.

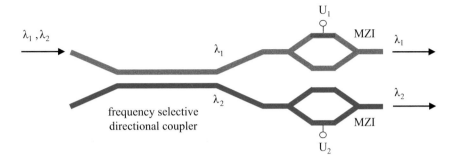

Fig. 8.5 1 × 2 switch with Mach-Zehnder interferometers (MZI)

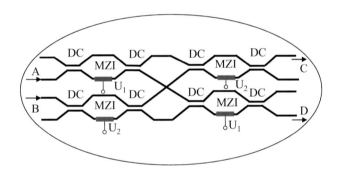

Fig. 8.6 2 × 2 Crossbar switch

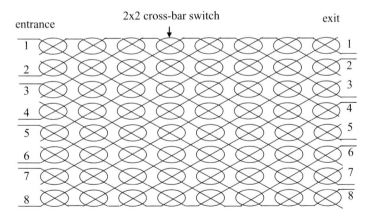

Fig. 8.7 8 × 8 switching matrix with 64 2 × 2 crossbar switches

Fig. 8.8 Controllable 1×2
X-switch

Another type of switch is the X-switch. In a 2×2 X-switch (Fig. 8.8), mode propagation in the crossing range of waveguide is influenced by the electro-optical effect. With the correct crossing range, we can obtain a constructive interference without voltage, resulting in the propagation of light from port 1 to port 4 (crossing state); with voltage, to port 3. With LiNbO$_3$ as electro-optical material one can realize an angle of intersection up to $3°$, resulting in centimeter dimensions of a single element (X-switch). Thus, cascading is possible.

Furthermore, in practice *digital switches* are in use (Fig. 8.9). One of the disadvantages of switches based on Pockels effect which we have talked about above is their critical dependence of the switching ratio (on-off) on polarization of light and on switching voltage. The switching behavior of digital switches, however, is different: there is a clear and stepwise switching behavior below or above a switching threshold, respectively.

In Fig. 8.9a, a Y-splitter is shown. Without voltage, modes will be split *equally* into *both* wings. Applying voltage, the refractive index will be increased by electro-optical effect. Thus, modes *cannot* propagate anymore, and the mode transfer runs completely in

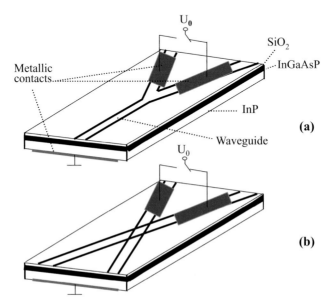

Fig. 8.9 Digital 1×2 Y-switch (a) and 2×2 X-switch (b)

the idle wing. If we apply voltage to the other wing signal is switched to the other port. A switching time of a few nanoseconds can be reached.

In digital X-switches (Fig. 8.9b) the basic idea is the same. Both X- and Y-switches can be arranged, e.g., on an InP substrate where the electro-optical material consists of InGaAsP. It is covered by a SiO_2 layer. The waveguide structure marked in Fig. 8.9 runs *inside InGaAsP*.

8.1.3   Mechano-Optical Switches

The basic idea of mechano-optical switching of optical signals is depicted in Fig. 8.10. A collecting and a diverging lens are placed back-to-back. If both lenses are situated on the optical axis, then they compensate each other (Fig. 8.10b). If we shift the diverging lens upwards, the light beam is also deflected upwards (Fig. 8.10a); conversely, a downward shifted beam is deflected downwards (Fig. 8.10c).

For fast switching, it is essential how fast the diverging lens can be moved to any other position. Practically speaking, the light will be deflected by a diverging lens on an actuator (Fig. 8.11, published with permission of the Fraunhofer Institut für Angewandte Optik und Feinmechanik Jena). Another lens array focuses light into the fiber to be switched.

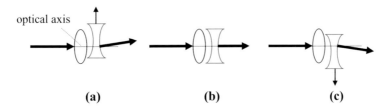

Fig. 8.10   Beam deflection by micro-optical lenses

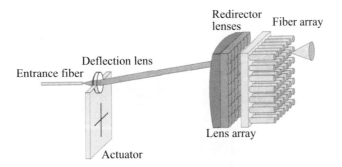

Fig. 8.11   1 × N faseroptischer Schalter

The free-space range between deflection lens and redirector lenses is about 20–50 mm long. Due to the small focal length of the micro lens, one is able to realize relatively large deflection angles (about 20°) even by a small shift (200–300 µm). The switching time is in the ms range. One problem is the relatively high voltage of about 150 V necessary for a piezoelectric shift.

8.1.4 Micro-Electromechanical Systems (MEMS)

MEMS (**M**icro-**E**lectro **M**echanical **S**ystems) is a mechanical switching by micro-mirrors. The basic idea is depicted in Fig. 8.12. A micro-mirror is controlled by a piezoelectric system which opens or closes an optical pathway.

Such a mirror (Fig. 8.13) has about 500 µm diameter. The gimbal-mounted mirror can be moved in two dimensions. The mirror is moved by piezoelectric controlled springs.

The mirror array MicroStar™ of T-Nova consists of 256 separately adjustable mirrors as depicted in Fig. 8.13 with a 1 mm mirror distance (Fig. 8.14a). Thus, we get a compact arrangement (Fig. 8.14b) with dimensions of about 25 mm x 50 mm x 50 mm. There are 256×256 channels which can be switched. The switching time is less than 5–10 ns, with crosstalk at about −50 dB.

In Fig. 8.15 a 4×4 switch based on two-dimensional MEMS is depicted.

Fig. 8.12 2×2 switch with MEMS

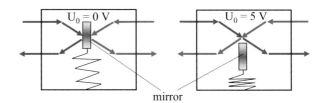

Fig. 8.13 MEMS of T-Nova (left) and lab photo of University of Applied Sciences Leipzig (right)

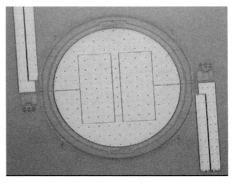

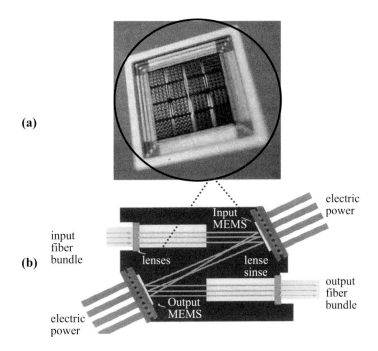

Fig. 8.14 MEMS array (**a**) and schema of a 3D-MEMS (**b**)

Fig. 8.15 4 × 4 Switch with
2D-MEMS

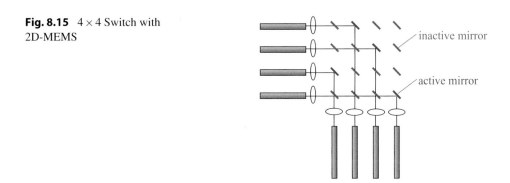

8.1.5   Thermal Switches

Due to the thermo-optical effect, the refractive index in quartz glass or plastics (poly-mers) changes with temperature. In quartz glass the change is approx. 10^{-5} per Kelvin, in polymers it is even greater. With a temperature change of approx. 100 K, refractive index changes can be achieved in a few milliseconds, which can lead to the switching of the light in the Mach-Zehnder interferometer or the Y-branch, for example. Today, inter-ferometric directional couplers and digital optical switches are commercially available for the second and third optical windows as 1×2, 1×4, 1×8 and 2×2 polymer-based switches. Their relatively low insertion loss of approx. 3 dB is also advantageous. About 10 to 20 W are required for control.

8.2   Filter in Optical Networks

The operation of a network with many channels (wavelengths with fixed bandwidths) requires the coupling or decoupling of closely spaced wavelengths (sometimes less than 0.3 nm). Optical filters serve this purpose.

8.2.1   Interference Filters (Thin-Layer Filter)

In thin-layer filters (Fig. 8.16a), the wavelength to be filtered is transmitted, and all others are reflected. The filter bandwidth is a few nm (Fig. 8.16b). Figure 8.16c shows a three-stage thin-film filter for use as a demultiplexer.

   In a waveguide interference filter, a waveguide structure in the planar technique is used [Mah 95]. In the waveguide interference filter (Fig. 8.17) several wavelengths are coupled at port 1 (in the figure two wavelengths λ_1 and λ_2). In a 3 dB coupler (see chapter 4) the total power is divided into 2 equal parts. The light passes through two interference arms of different lengths (wings 1 and 2 in Fig. 8.17), after which both parts interfere in a frequency-selective directional coupler. Consequently, one wavelength (λ_1) exits from port 3 and the other (λ_2) from port 4. The frequency-selective directional

Fig. 8.16 Thin film filter (a), transmission curve (b), and application example as demultiplexer (c)

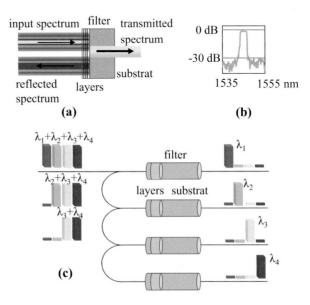

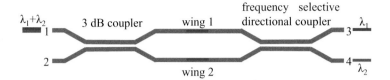

Fig. 8.17 Waveguide interference filter

coupler is precisely tuned to *one specific wavelength* (here λ_1). If there are more than two wavelengths, only *one wavelength* is coupled out at port 3, while all others occur together at port 4 - thus a cascading of the waveguide interference filter is necessary. Problematic for waveguide interference filters are the requirements for the frequency-selective coupler (setup for *one wavelength* with effects on the crosstalk), the temperature stability, and the high requirements for polarization and polarization maintenance. This results in a typical channel spacing of 0.1–5 nm and a crosstalk attenuation of less than -20 dB, for cascaded arrays 0.8–5 nm and less than -15 dB is achieved.

For such a planar waveguide structure, decoupling from and recoupling into the optical fiber (transition from a circular core cross section to a rectangular waveguide or vice versa) is always problematic. This is always associated with losses.

Figure 8.18a shows a transmission and reflection grating in which the individual wavelength components are separated by diffraction. With a lens, a demultiplexer can be realized with a reflection grating (Fig. 8.18b).

8.2.2   Fabry-Perot Filters

If one interrupts the fiber and positions the end faces of the fiber parallel to each other with a distance of a few μm, then one gets a Fabry-Perot filter. In this case we use the wavelength-selective properties of two vis-à-vis positioned reflecting surfaces (Fig. 8.19). A *certain distance* between mirror-like surfaces corresponds to a *certain wavelength* (in Fig. 8.19 for wavelength λ_g). We get standing waves. Thus wavelength λ_g is reflected and can be decoupled by a frequency-selective coupler (not depicted in

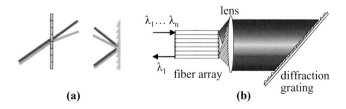

(a)   **(b)**

Fig. 8.18 Transmission or reflection diffraction gratings (a) and their application as demultiplexers (b)

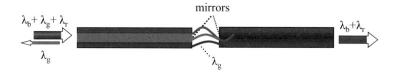

Fig. 8.19 Fabry-Perot-Filter

Fig. 8.19). All the other wavelengths (in Fig. 8.19 "red" light with λ_r and "blue" light with λ_b) are transmitted and propagate still in the fiber. The mirror distance (half the wavelength would be perfect), quality of end faces (surface roughness), and reflectivity of mirrors determine the channel distance (0.1–5 nm are typical values) and the crosstalk (typically less than -17 dB).

8.2.3 Fiber-Bragg-Gratings

In a **F**iber-**B**ragg **g**rating (FBG), inside the fiber a grating-like periodic structure of refractive index is created by dopants (i.e., refractive index alternates between n_H and n_L, see Fig. 8.20). This periodic structure acts like feedback in a DFB laser (see Sect. 5.3.2). Light of a wavelength exactly "matched" to the grating (i.e., half wave is equal to grating period d) is reflected (green wave with λ_g in Fig. 8.20). All the other wavelengths (Fig. 8.20 "red" light with λ_r and "blue" light with λ_b) are *not* reflected. Grating period, difference of refractive index between "high" (n_H) and "low" (n_L) as well as the number of gratings determine the channel distance and the crosstalk. Typical values are 0.5–5 nm and less than -30 dB, respectively. Because the fiber is not artificially interrupted, the insertion losses are low.

8.2.4 Phased-Array Routers

Waveguide grating routers (WGR) or arrayed waveguides (AWG) are designed as planar waveguides. The substrate is mostly mono-crystalline silicon (Si) or indium phosphide (InP), sometimes also glass. Wave guiding structures are fabricated by doping the substrate.

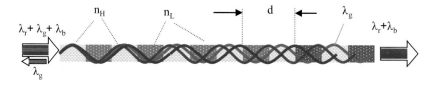

Fig. 8.20 Mode of operation of Fiber-Bragg grating

In AWGs, the first light of different wavelengths (in Fig. 8.21 three wavelengths $\lambda_1 + \lambda_2 + \lambda_3$) will be decoupled from the fiber into the waveguide structure. In the so-called free-space range (FSR), the light of all wavelengths will be partitioned evenly and as low loss as possible to waveguide structures (for simplicity in Fig. 8.21 only three waveguides are indicated).

From the different lengths of the waveguides, different pathways result (the difference between neighboring waveguides is ΔL). This means that at the end of the waveguide, we have different phases. In a second free-space range, light from the waveguides can interfere and we get higher-order diffraction. Thus, we get a higher-order interference pattern; and wavelengths λ_1, λ_2, and λ_3 are separated in space and can be coupled separately into outgoing fibers. In Fig. 8.21 demultiplexing by AWG is described—for wavelength multiplexing we have to "invert" the optical pathways (one can consider light from right to left in Fig. 8.21). In AWG we have a low channel distance (0.1–2 nm), and crosstalk is less than -15 dB. Typical problems are losses during coupling and decoupling. In Fig. 8.22 one can see practical examples.

8.2.5   Application of Optical Filters

8.2.5.1   Optical Isolator

An optical isolator, or Faraday rotator, is an optical valve which permits light to pass only in one direction. Propagation of ray 1 in Fig. 8.23a through the optical isolator is nearly undisturbed (insertion losses < 1 dB), whereas ray 2 is blocked (typical attenuation about 70 dB). Optical isolation is based on magneto optical or the Faraday Effect. Under the influence of a permanent magnetic field of field strength H, polarization of light is rotated. The rotation angle φ is given by

$$\varphi = V \cdot L \cdot H \tag{8.4}$$

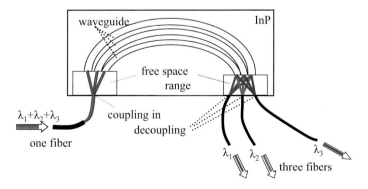

Fig. 8.21   Function of an arrayed waveguide (AWG)

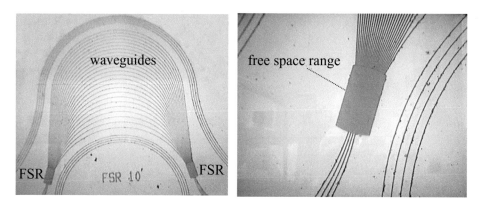

Fig. 8.22 AWG: practical example (author's lab pictures)

with V—Verdet constant and l—interaction length. Due to its high magneto optical effect, often crystalline Yttrium-Iron-Garnet (YIG) is used.

In the direction of polarizer 1, polarized light (Fig. 8.23b) will have a polarization rotation of 45° in YIG. Polarizer 2 is placed exactly under 45°; the ray can pass without any losses. If ray 1 is completely or partly reflected (ray 2), it can pass polarizer 2 without losses. After that, in YIG its polarization is rotated at 45° - and now it is blocked by polarizer 1.

In near infrared (NIR), terbium-doped glass or Terbium-Gallium-Garnet ($Tb_3Ga_5O_{12}$, abbreviated TGG) is used. At wavelength of more than 1.1 µm one can also use Yttrium-Iron-Garnet ($Y_3Fe_5O_{12}$, abbreviated YIG). The necessary magnetic field can be generated by strong permanent magnets based, for example, on a mixture of neodymium, iron, and boron; with these elements, the length of the optical isolator can be kept relatively short (one or some centimeter).

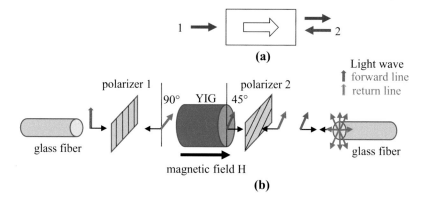

Fig. 8.23 Optical isolator: Block diagram (a) and setup (b)

8.2.5.2 Optical Circulator

Optical circulator is an interconnect with three or more ports connected by unidirectional links pointing in the same direction. In an optical circulator light runs between ports in a loop. A typical block diagram can be seen in Fig. 8.24a. Up to six ports can be used. Insertion loss is about 0.5–1 dB.

Light propagation from port 1 to port 2 is sketched in Fig. 8.24b:

- In a birefringent walk-off block BWOB A, unpolarized light from port 1 will be separated in an ordinary (−•−•−•−) and extraordinary (+++) polarized light and separated in space.
- In a Faraday rotator, polarization is rotated at 45° *clockwise*.
- In a phase retarder (λ/4-plate), polarization is rotated again at 45° *clockwise*, thus the total polarization rotation is 90°.
- In a birefringent walk-off block BWOB B, ordinary and extraordinary polarized light are combined; now it leaves the circulator at port 2.

Light propagation from port 2 to port 3 is sketched in Fig. 8.24c:

Fig. 8.24 Optical circulator: Block diagram (a) and setup of wings $1 \rightarrow 2$ (b) and $2 \rightarrow 3$ (c)

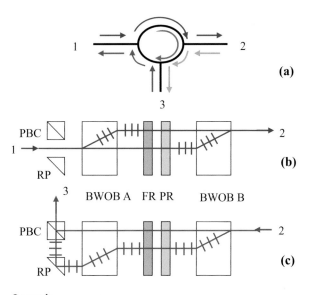

Legend:

PBC	– polarizing beam splitter cube
RP	– reflector prism
BWOB	– birefringent walk-off block
FR	– Faraday rotator
PR	– phase retarder

- In a birefringent walk-off block BWOB B, unpolarized light from port 2 will be separated in an ordinary and extraordinary polarized light and separated in space.
- In a phase retarder, polarization is rotated again at 45° *anti-clockwise.*
- In a Faraday rotator, polarization is rotated at 45° *clockwise*, thus the total polarization rotation is 0°.
- In a birefringent walk-off block BWOB B, ordinary and extraordinary polarized light will be (again) separated and combined by a reflector prism and a polarizing beam splitter cube; then the light leaves the circulator at port 3.

Light propagation from port 3 to port 1 runs like from port 2 to port 3.

8.2.5.3 Optical Add-Drop Multiplexer

In an Add-Drop-Multiplexer (Fig. 8.25) from the many data channels (wavelengths λ_1, λ_2, λ_3 and λ_4 in Fig. 8.25) coming via port 1, *one wavelength* (λ_3) which came over glass fibers to port 1 will be decoupled to port 4 (drop). At the same time, a data channel on the now freed-up wavelength λ_3 in Fig. 8.25 will be coupled into port 2 (add). Thus, one *configuration by data channels* (wavelengths) is changed, but the wavelengths are *unchanged*. In an Add-Drop-Multiplexer, decoupling and coupling are performed exclusively optically using couplers and switches.

As an example, an 8-channel Add-Drop-Multiplexer with 16 MEMS is shown in Fig. 8.26. We have 8 data channels (wavelengths) *in one fiber*. First, we have to perform a demultiplexing to separate these 8 channels. By micro-mirrors *one wavelength* will be decoupled (drop). Now this channel is free, and a "new" data channel can be coupled in (add). After multiplexing we again have 8 channels in one glass fiber.

There are 2 versions of optical Add-Drop-Multiplexers:

- static OADM (fixed OADM or FOADM, see Fig. 8.27a)
- flexible, externally configurable OADM (remote OADM or ROADM, see Fig. 8.27b)

At FOADM we have a static configuration, i.e., there is a fixed node at the entrance and the exit for each wavelength. It can be changed only by manual operation. In contrast, in ROADM channels to be decoupled and/or coupled can be chosen freely and flexible. To configure ROADM, one needs local transmitters and receivers and a routing system

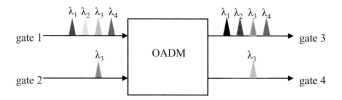

Fig. 8.25 Add-Drop-Multiplexer

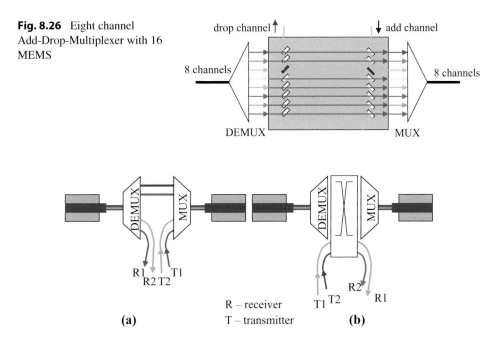

Fig. 8.26 Eight channel Add-Drop-Multiplexer with 16 MEMS

Fig. 8.27 FOADM (a) and ROADM (b)

which is able to switch optionally input signals to the output or to a local receiver. A ROADM can be controlled from a distance (remote control).

8.2.5.4 Optical Cross-Connector

An Optical Cross-Connector (OXC) is mainly used to adapt the *data amount* (bit rates) per glass fiber, and accordingly to optimize the data transfer in a glass fiber network.

First light from N glass fibers (Fig. 8.28) will be demultiplexed separately for each fiber—thus we get N·n data channels (n is the number of wavelengths transferred *in a single fiber*). After that, in a *space step,* light from one fiber is coupled to another one by a "switch-over".

In Fig. 8.28 light of wavelength λ_{11} from fiber 1 is coupled into fiber M (at wavelength λ_{M5}), whereas light of wavelength λ_{N5} from fiber N is coupled into fiber 1 (wavelength λ_{11}). This *space step* can be realized, e.g., by three-dimensional MEMS (see Fig. 8.14b), where light from one fiber is coupled to another fiber by means of two movable mirrors (3D-MEMS).

Maybe the data channel of wavelength λ_{M5} in fiber M is "occupied" - in this case it is "blocked". Now we have to change the wavelength of data channel in a *frequency stage* (Fig. 8.28) from λ_{15} to λ_{11} - if this channel is free!

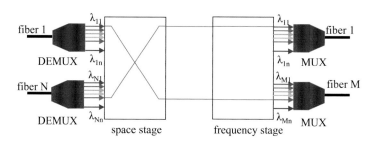

Fig. 8.28   Function of an optical Cross-Connector

This frequency step is a real problem for all-optical networks - in optics, it is only possible to realize by nonlinear effects, such as the **four-wave m**ixing (FWM), but this requires very high power (see Chap. 10).

8.3   Signal Regeneration

The task of a regenerator is to completely restore the signal so that it has the same form as the input signal. The noisy optical signal is detected, converted into an electrical one and electronically processed. Partial regeneration can also be purely optical. Complete signal regeneration in the optical plane is the subject of research by numerous research groups.

There are three stages of signal regeneration (Fig. 8.30):

1. Amplification (Re-Amplification)
2. Clock recovery (Re-Timing)
3. Pulse shaping (Re-Shaping)

1R optical regeneration is performed using optical amplification and is state of the art. Re-shaping can also be found in the form of dispersion compensation. "All-Optical

Fig. 8.29   Space stage of an 8 × 8 Cross-Connector with 3D-MEMS

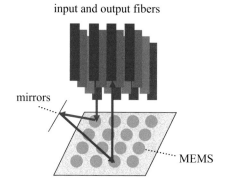

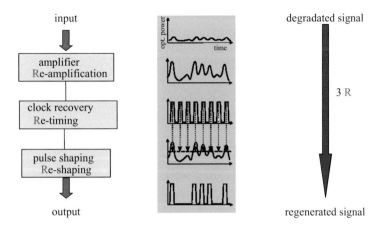

Fig. 8.30   3R regeneration

Regeneration" is very difficult due to the difficulties in clock recovery and is not yet commercially available.

For this reason, the purely optical transmission path is still interrupted in practice today and transponders (Fig. 8.31) are used in which a receiver (R) first converts the data into an electrical signal (opto-electronic conversion), which is then used to modulate the light of a laser with a suitable wavelength (electro-optical conversion). The laser diode LD is operated by a bias current I0 and modulated by a data current ID. The pure wavelength conversion in the transponder is usually combined with an amplification of the optical signal; this is referred to as a 1R transponder (amplification only, Re-amplification). Of course, the bit shape can be changed at the same time (e.g., dispersion compensation, re-shaping) or the temporal position of the bit can be corrected (re-timing), in which case we speak of a 3R transponder (Re-shaping, Re-timing, Re-amplification). In Fig. 8.31, S stands for Re-shaping, T for Re-timing, and A for Re-amplification. In each case, the purely optical transmission path is interrupted by opto-electronic-optical conversion; we speak of semi-transparent or opaque networks.

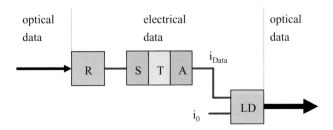

Fig. 8.31   Function of a 3R transponder

However, since the cost of such transponders is not inconsiderable, attempts are being made to avoid frequency conversion and to make do with a spatial stage alone (as shown, for example, in Fig. 8.29).

For purely optical networks, optical amplifiers are required above all.

8.4   Optical Amplifier (Re-Amplification)

Optical amplifiers have the exclusive task of increasing the light output (power amplification), whereby all other parameters (bit length, line width, noise, etc.) should remain as unaffected as possible. The shape of the bits and their position in time (time shift) are generally not changed by the amplifier. The aim is complete integration of the amplifier into the optical network, i.e., no conversion of optical signals into electrical signals or vice versa is permitted.

The increase in level (gain) is measured in decibels (dB), as always in optical communications technology, where the relationship to the power gain is given by

$$g^{dB} = 10 \cdot \log \left(\frac{P_2^{mW}}{P_1^{mW}} \right) \tag{8.5}$$

where P_1^{mW} or P_2^{mW} is the optical power (in mW) before or after amplification, respectively. Consequently, a doubling of power means a gain of +3 dB.

Today and in near future three types of optical amplifiers are commercially used:

- Erbium-doped fiber amplifier (EDFA).
- Raman optical amplifier and
- Semiconductor optical amplifier.

In the following we will discuss function, advantages, and disadvantages as well as problems using these amplifiers.

8.4.1   Erbium-Doped Fiber Amplifier (EDFA)

8.4.1.1 Amplification in Erbium-Doped Glass Fibers

The erbium-doped glass fiber amplifier (EDFA) consists of a specially prepared by Erbium ion dopants glass fiber of about 10–50 m length depending on concentration of dopants.

The energetic structure of rare earth erbium (Er^{3+}) is depicted in Fig. 8.32. From many energetic bands, only the upper (completely occupied) band E_1 and (nearly empty) bands E_2 and E_3 are of importance. The distance between E_2 and E_1 corresponds to the Third optical window around 1.55 µm. Like in a laser we can get amplification if in band

Fig. 8.32 Energetic structure of Er^{3+} doped silica glass

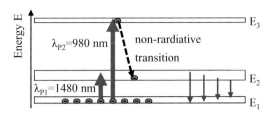

E_2 should be more electrons than in band E_1. We can get it by optical pumping by a laser emitting light in the absorption range of the EDFA.

Spectrum of absorption and emission (in arbitrary units) is depicted in Fig. 8.33. We can achieve especially large absorption we get at $\lambda_{P1} = 980$ nm and $\lambda_{P2} = 1480$ nm. This pump process brings electrons from E_1 to E_2 (1480 nm) or from E_1 to E_3 (980 nm), in this last case electrons recombine by non-radiative transition of electrons to E_2 (Fig. 8.32). An absorption range of about 850 nm is not useful for pumping. To have more electrons in the energy band E_2 (so-called inversion) it is necessary to repeat the pump process sketched in Fig. 8.32 very often. Thus, we need high pump energy (100 mW or more).

Electrons of energy band E_2 recombine statistically back to energy band E_1 (*spontaneous* emission). If we have an inversion by optical pumping, light to be amplified will be amplified by *stimulated* emission. Signal amplification we can get in the amplification range (emission line width) between 1525 and 1565 nm (Fig. 8.34). Thus, using **one** EDFA, we are able to amplify different wavelengths close to each other. This is illustrated in Fig. 8.32 by arrows of different lengths. It results in the amplification spectrum of Fig. 8.34.

Of course, this amplification spectrum depends on the pump power (Fig. 8.35). Modified EDFAs which are also suitable for the L band (1570–1610 nm) now exist.

As one can see in Fig. 8.35 we get amplification only above a certain pump power (about 10 mW in Fig. 8.35). This threshold power is necessary to overcome the attenuation in the glass fiber. The gain efficiency (relation between gain and pump power) is maximum in point 1. Further increase of pump power results in saturation of amplification. Furthermore, we have a dependence of gain g on the length of the Er^{3+} doped glass

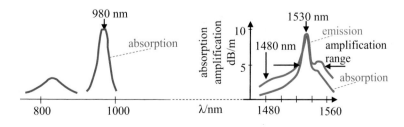

Fig. 8.33 Absorption (red) and emission spectra (blue) of Er^{3+}-ion doped glass fiber

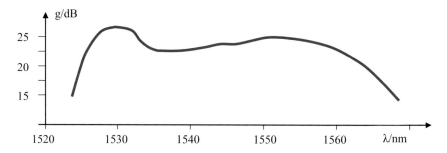

Fig. 8.34 Amplification spectrum of Er^{3+} doped glass fibers

Fig. 8.35 Dependence of amplification on pump power

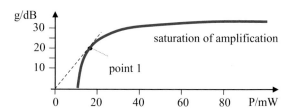

fiber L. Power decreases along the fiber exponentially by the (desired) absorption. Thus, after a certain fiber length, the total amplification does not increase any more—above an optimum length the fiber seems to be no longer "necessary", because gain by amplification is less than losses by absorption.

This dependence can be simulated using the following assumptions (see Fig. 8.36):

- Pump power decreases exponentially due to absorption of pump light in the fiber. Therefore, the amplification g_L at a certain fiber length decreases also exponentially $g_L = P_0 \exp(-L)$ (red line in Fig. 8.36), P_0 is the pump power at the beginning of the EDFA.
- Absorption a is constant along the fiber (green line in Fig. 8.36).
- If one takes the difference between local amplification g_L and absorption a, and integrates over the length L, we get the total amplification g (blue lines in Fig. 8.36 where

Fig. 8.36 To modeling of amplification in an EDFA

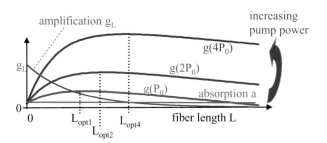

power P_0 is doubled for each solid line). The model calculation can be reproduced using a MathCad program (EDFA in the "Extras" folder) (can also be viewed as a pdf file).

Thus, the optimum length L_{opt} is different for each pump power (Fig. 8.36). Therefore, to obtain optimum amplification in an EDFA of given length, one needs a certain (optimized) pump power.

Exercise 8.3

In which cases (wavelength range) can one not use EDFA?

▶ **Tip**
Help H8-3 (Sect. 12.1).
Solution S8-3 (Sect. 12.2).

8.4.1.2 Noise in EDFAs, ASE

Noise in amplifiers is characterized by the *noise factor*. Noise factor is the relation of **S**ignal-**N**oise **R**atio (SNR) between entrance (SNR_{in}) and exit (SNR_{out}) of amplifier:

$$F = \frac{SNR_e}{SNR_a} \tag{8.6}$$

In decibel, the noise factor F* is given by

$$F^* = 10 \cdot \log(F) \tag{8.7}$$

The signal-noise ratio is the ratio of signal P_S to noise power P_n. In a noiseless amplifier, the signal-noise ratio is the same for exit and entrance. Thus, we get $F = 1$ or $F^* = 0$ dB, respectively.

The main reason for noise in erbium-doped glass fibers is spontaneous emission of light. This spontaneous emission will be amplified during propagation along the fiber. Thus, we get the **a**mplified **s**pontaneous **e**mission (ASE). For ASE we get the same amplification spectrum as depicted in Fig. 8.34. However, the intensity of ASE is much higher *without* the signal, when compared to the intensity which occurs *with* signal (Fig. 8.37a). The reason is that for amplification (transitions from E_2 to E_1) the upper energy level E_2 is "cleared", so that generation of noise by ASE is more difficult. Without the signal, the noise power can be close to 0 dBm (1 mW), but with the signal it is decreased down to about -12 dBm. Thus, signal can be amplified by more than 20 dB (Fig. 8.37a). Noise factor depends on the signal power p_S at the entrance; we can find a minimum at about $p_S \cong -20$ dBm (Fig. 8.37b). Furthermore, the noise factor depends on the pump wavelength; at $\lambda_p = 980$ nm we find the theoretically possible minimum of 3 dB.

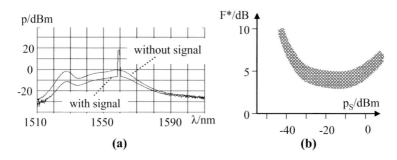

Fig. 8.37 Amplification spectrum at EDFA exit (a) without or with a signal at entrance and dependence of the noise factor on signal at entrance (b)

To reduce the influence of ASE in EDFA, in at least one channel (at one wavelength) we should have a 1-Bit. Furthermore we should keep in mind that, at 22 dB amplification *in one channel* (in Fig. 8.38 at $\lambda = 1559$ nm), we can find more or similar ASE power P_{ASE} at other wavelengths (in Fig. 8.38 at about 1528 nm).

Using an optical filter (if possible, one which can be controlled), one can reduce ASE power in the wavelength range not in use. Using this filter, one can reduce noise in general.

8.4.1.3 Dense Wavelength Division Multiplexing (DWDM) and EDFA

Using DWDM (see also Chap. 10.2.3) we have many data channels close to each other with a channel distance of 50 GHz (corresponding to about 0.4 nm in the Third optical window) or less [Brü 01]. Amplification in Er^{3+} doped glass (Fig. 8.39, after [Gla 97] to a great extent depends on the wavelength of the data channel in the Third optical window (Fig. 8.39a). As one can see in Fig. 8.39a after quadruple amplification, we get big differences in the power of amplified channels (up to about 20 dB). On the other hand, further into the pass in the fiber, the attenuation is the same for all data channels. Thus, the distance to the next EDFA is determined by the least channel. Using many amplification cascades, we can expect very large differences in signal power of different data channels. To avoid this, we can start with an inverse power spectrum from transmitters matched to the amplification spectrum (more power at low amplification and verse vice). Of course, this method is suitable only if the length of the optical transmission length (= fiber length) is known.

Fig. 8.38 Signal together with noise spectrum at exit of an EDFA

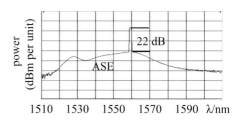

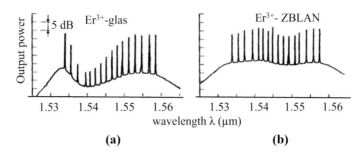

Fig. 8.39 Spectrum of 16 channels after quadruple amplifier pass through Er^{3+} doped glass (a) and Er^{3+} doped ZBLAN (b)

If, instead of silica glass, one uses a special (without silica) glass ZBLAN (abbreviation for a mixture of ZrF_4, BaF_2, LaF_3, AlF_3 and NaF) we get a smoother and, therefore, broader emission spectrum (Fig. 8.39b).

Exercise 8.4

Which problems result from different amplification (some dB) of neighboring channels?

▶ **Tip**
Help H8-4 (Sect. 12.1).
Solution S8-4 (Sect. 12.2).

Exercise 8.5

How can we get a smoothing of amplification spectrum?

▶ **Tip**
Help H8-5 (Sect. 12.1).
Solution S8-5 (Sect. 12.2).

8.4.1.4 Experimental Realization of EDFAs

Some peculiarities and possibilities of an experimental setup of EDFA should be mentioned. By a coupler, pump power will be coupled into the erbium-doped fiber. To avoid any back reflection of the signal light by the amplifier, one can use optical isolators (also called a Faraday rotator, see Sect. 8.2.5.1). A Faraday rotator acts like a valve which avoids the backward reflection of light in the direction of the transmitter.

With optical pumping in a *forward direction* (Fig. 8.40a), we can expect high amplification at the beginning of EDFA. Then the amplification is reduced with increasing fiber length. Thus, pumping in a forward direction can be used predominantly for low-noise amplification of very weak signals.

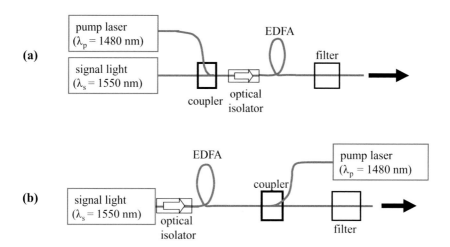

Fig. 8.40 Optical pumping in forward (a) and backward (b) directions

With optical pumping from a *backward direction* (Fig. 8.40b), we have the highest amplification at the end of EDFA. With a decreasing influence of pumping, amplification is reduced in the direction of the beginning of the fiber. Thus, pumping from a backward direction can be used predominantly for amplification of relatively strong signals at the beginning of EDFA; in this case we can reach amplification saturation (up to +18 dBm).

Both versions can be combined (Fig. 8.41). Pumping *in forward and from backward directions,* we are able to combine the advantages of both versions—amplification of weak signals up to amplification saturation.

Thus, three typical applications of EDFAs are possible:

- As *pre-amplifier* to get low-noise amplification of very weak signals just in front of the receiver - to this end we can realize optical pumping in forward direction.
- As *intermediate amplifier*, in this case we expect both amplification of weak signals and high output power - this is the strength of forward *and* backward pumping.

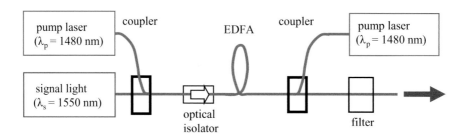

Fig. 8.41 Pumping from both sides (forward and backward direction)

- As *final amplifier*, especially to increase the transmitter power up to about +20 dBm (100 mW) - in this case pumping from backward direction is of advantage. Of course, in this case we have to accept more noise.

Exercise 8.6

In which case is pumping in the forward direction of advantage, and in which case is pumping from the backward direction of advantage?

▶ **Tip**
Help H8-6 (Sect. 12.1).
Solution S8-6 (Sect. 12.2).

8.4.1.5 Other Amplifiers Doped with Rare Earths

In addition to Erbium ions, one can also use ions of other rare earths as amplifier (e.g., Eq. 3.1). In the near future, especially praseodymium (Pr^{3+}) can be expected as a commercially available amplifier for the second optical window (Table 8.1).

8.4.2   Raman Amplifier

8.4.2.1 Raman Effect

Glass, or SiO_2 is a three-atomic molecule arranged without any structure (amorphous). One can imagine this molecule as two oxygen atoms "coupled" by "springs" to silicon (model of dumbbells, see Fig. 3.14). In such a molecule we have vibrations. From theory, we know that in a molecule like SiO_2 there are three different basic vibrations, which are sketched in Fig. 8.42. In Fig. 8.42a we have the vibrations between oxygen and silicon atoms (stretch vibration, frequency f_1), in Fig. 8.42b we have vibrations between oxygen atoms (swing vibration, frequency f_2) and in Fig. 8.42c oxygen atoms vibrate together (twist vibration, frequency f_3). Each vibration represents a certain energy $h \cdot f_v$ (subscript v stands for the word "vibration"), which should be added to the (existing) electronic energy level. These vibration frequencies are in the IR of the spectrum.

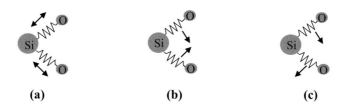

| (a) | (b) | (c) |

Fig. 8.42 Types of basic vibrations in SiO_2

The Raman Effect was discovered in 1928 by the Indian physicists Chandrasekhara Venkata Raman (1888–1970) and Kariamanikkam Srinivasa Krishan (1898–1961); for that discovery only Raman received the Nobel prize for physics in 1930.

Raman scattering is the inelastic scattering of photons in matter. If light waves (photons) with the photon energy $h \cdot f$ impact to vibratory molecules (Fig. 8.43), at the inelastic scattering a certain part of the photon energy is transferred to the vibration, This then results in changes of frequency or wavelength of light (Fig. 8.42).

In Fig. 8.43 the energy levels are depicted. The lowest (completely occupied with electrons) energy level is determined by electron configuration (electronic level); above that level we have the first vibration level (given by one of the vibrations of Fig. 8.42). Photon energy $h \cdot f$ of the incident light corresponds to an arrow. Its length corresponds to the distance between electronic and virtual (i.e., not existing) energy levels.

As a result of Raman scattering, part of the photon energy is transferred to the vibrational frequency f_v of the molecule, so the frequency of the scattered light is now f_S, where $f_S = f - f_v$ is called the Stokes frequency. Light with the Stokes frequency f_S is emitted (Raman scattering), and the electron remains in the excited state at the vibrational level $E_v = h \cdot f_v$ (Fig. 8.43a). This is the so-called *spontaneous Stokes Raman scattering*, i.e., the light of the Stokes frequency is emitted randomly in time and in all directions. The corresponding energy balance of spontaneous Raman scattering can be written as

$$h \cdot f_S = h \cdot f_P - h \cdot f_v \tag{8.8}$$

And vice versus, if the incident light meets vibrating molecules, the vibration energy will be added and scattered light has the frequency $f_{AS} = f + f_v$ (Anti-Stokes shift, Fig. 8.43b), this is also *spontaneous Anti-Stokes Raman scattering*.

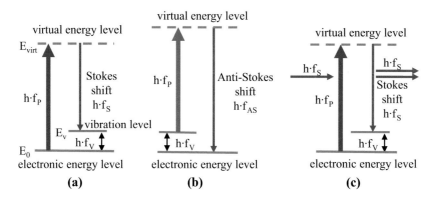

Fig. 8.43 Raman effect: Spontaneous Stokes (a) and Anti-Stokes (b) Raman scattering as well as stimulated Raman scattering (c)

However, if there is already some light (e.g., from spontaneous Raman scattering) with the Stokes frequency f_S in a certain direction, one can obtain an enhancement by stimulated Raman scattering (Fig. 8.43c). The energy balance of stimulated Raman scattering can then be written as follows:

$$h \cdot f_S = h \cdot f_P - h \cdot f_P + h \cdot f_S \tag{8.9}$$

Thus, it is a nonlinear four-particle effect (see Chap. 10), two particles of pump frequency f_P, one particle of spontaneous Raman scattering of frequency f_S and one stimulated particle of Raman scattering of frequency f_S. This is the Raman optical amplifier (ROA).

8.4.2.2 Spontaneous Raman Scattering, Spectrum of Amplification

The Raman process described so far is spontaneous, i.e., the resulting Stokes or anti-Stokes waves are not in phase (spontaneous Raman scattering), moreover, the Raman scattering goes in all directions in space. So far, we have considered only the frequencies f or energies h·f (law of conservation of energy). If we additionally consider the momentum or phase of the wave (momentum conservation law), we can see that Stokes waves propagate parallel to the pump wave, but the anti-Stokes waves propagate conically around the pump wave. Therefore, only the Stokes waves are of interest for a Raman amplifier.

During the pumping process (absorption of the pump radiation), amplified spontaneous scattering or emission (ASE) occurs. The corresponding power P_{ASE} (in W) depends on the optical bandwidth of the signal Δf_S, the equivalent noise factor F_n and the gain $G = \frac{P_{S.an}}{P_{S.aus}}$ (dimensionless) and can be described by the following expression [F. Cisternino and B. Sordo: State of the art and prospects for Raman amplification in long-distance optical transmissions. Telecom Italia Lab. Journal exp 2, No. 1, 18–25, 2002]:

$$P_{ASE} = h \cdot f_S \cdot \Delta f_S \cdot F_n \cdot (G - 1)$$

with Planck's constant $h = 6.626 \cdot 10^{-34}$ Js.

The power of the ASE can be estimated with the following quantities: Bandwidth $\Delta f_S = 25$ GHz (corresponding to $\Delta\lambda = 0.2$ nm), signal frequency $f_S = 1.935 \cdot 10^{14}$ Hz (corresponding to $\lambda_S = 1.550$ nm), noise factor $F_n = 4$ and the gain $G = 20$ (corresponding to a gain in dB of 13 dB). Using these figures, the power of the amplified spontaneous emission is obtained to be $P_{ASE} = 2.436.10 - 4$ W.

Figure 8.43 gives only one Stokes frequency f_S, in reality one is dealing with broadened lines—the Raman gain spectrum (Fig. 8.44) is obtained as the frequency or wavelength range in which the emission is greater than the absorption. The maximum of the Stokes shift is then obtained at the Stokes frequency f_S.

The maximum Stokes shift in pure fused silica (SiO_2) and in vitreous germanium glass (GeO_2) is at wavenumber 440 and 420 cm^{-1} (corresponding to frequency $f_S = 13$ THz and a shift of about 100 nm in the region of the third optical window) and $f_S = 12$ THz, respectively. The gain line width (3 dB drop) of the Raman amplifier is about

Fig. 8.44 Raman gain
spectrum for pumping with
$\lambda = 1450$ nm

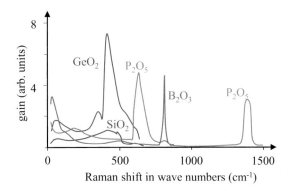

$\Delta f_R = 7$ THz (about 40 nm). As can be seen in Fig. 8.44, the maximum of Raman intensity in GeO_2 at 420 cm^{-1} is about 7.4 times higher than in pure SiO_2.

The maximum of the shift in vitreous B_2O_3 is about 808 cm^1. The Raman intensity
is about 4.6 times higher than in pure SiO_2. Glassy P_2O_5 has a maximum at 640 cm^{-1}
(Raman intensity 4.9 times higher) and another at 1390 cm^{-1} (Raman intensity 3.0 times
higher).

The parameters of some optical fibers such as maximum Raman gain $C_{R,max}$ in
W^{-1} km^{-1}, absorption coefficient α in dB/km for different wavelengths, dispersion D in
ps/km · nm ps/km · nm at $\lambda = 1550$ nm and the dispersion slope (slop) S can be obtained
from the data sheets, they are shown in (e.g., Eq. 3.1). The absorption k is given in km^{-1},
where the relation with the attenuation coating α (in dB/km) is given as

$$k\left(\text{in } km^{-1}\right) = \frac{\alpha\left(\text{in } \frac{dB}{km}\right)}{4.343}$$

For example, for the signal wave of $\lambda = 1550$ nm with $\alpha_S = 0.2$ dB/km the value
$k_S = 0.0463$ km^{-1} is obtained, for pump wave of $\lambda = 1450$ nm with $\alpha_P = 0.25$ dB/km the
value $k_P = 0.0575$ km^{-1} is obtained. Parameters of various glasses are given in Table 8.2.

Figure 8.45 shows the Raman coefficient C_R as a function of Stokes shift for different
glasses. The following legends were used: 1 - Standard SMF (SSMF), 2 - True Wave™
REACH (TW-REACH), 3 - True Wave™ RS (TW-RS), 4 - dispersion compensating
fibers (DCF) for SSMF, 5 - DCF for TW-RS. In addition, the location of a 21-channel
transmission system is noted in Fig. 8.45, with the "blue side" (higher frequencies) on
the right and the "red side" (lower frequencies) on the left. For comparison, the line
shape (Lorentz line shape) is also shown, which can be calculated from the following
formula:

$$L(f) = \frac{\left(\frac{\Delta f}{2}\right)^2}{\left(\frac{\Delta f}{2}\right)^2 + (f - f_0)^2}$$

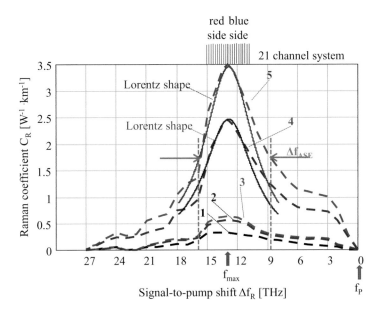

Fig. 8.45 Raman coefficient C_R for various vitreous materials

where Δf is the line width and f_0 is the mean frequency. A comparison of the linewidths (Gaussian or Lorentzian shaped) was performed in MathCad, the file is listed in the folder "Extras" under "Lorentz Gauss" (can also be seen as pdf file). The spectra shown are the amplified spontaneous Raman emission (ASE) with linewidth Δf_{ASE} (falling to half the value of C_R, see Fig. 8.45). The value of Δf_{ASE} is approximately the same for all vitreous materials and is about 7.2 THz.

Because the Stokes shift is fixed to the pump wavelength, a shift in the Raman gain spectrum can be achieved by changing the pump wavelength. In contrast, the pump wavelength and the gain range are fixed in the EDFA—variation is not possible.

Obviously, the maximum gain depends linearly on the pump power. From Fig. 8.46 it can be seen for SiO_2, for example, that a doubling of the pump power from 100 to 200 mW results in about twice as much Raman gain (practically about 4 dB), while the spectral shape as well as the gain bandwidth $\Delta \lambda_{ASE}$ is preserved.

The use of 2 or more different pump wavelengths offers the possibility to extend and/or smooth the gain spectrum (Fig. 8.47). A "smooth" gain spectrum offers decisive advantages especially for DWDM (see also discussion on Fig. 8.39). There is a limit to the power; 2 pump lasers of 27 dBm and 3 pump lasers of 28.5 dBm are reported, with gains of 10 and 14 dB, respectively. It should also be noted that, in accordance with an ITU specification, the total power in the optical fiber must not exceed 30 dBm at any time or place because of the risk of destruction.

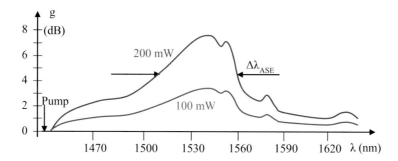

Fig. 8.46 Gain spectrum of spontaneous Raman scattering in SiO_2 at 100 and 200 mW pump power, respectively

---

Exercise 8.7

A still broader and smoother amplification spectrum can be obtained using three or more pump sources. How is this method limited?

▶ **Tip**
Help H8-7 (Sect. 12.1).
Solution S8-7 (Sect. 12.2).

8.4.2.3 Stimulated Raman Scattering

If the incident light of frequency f_p is considered as pump light and if the signal light of frequency f_S to be amplified is irradiated exactly in the direction of the pump light, induced (stimulated) Raman scattering can be forced, i.e., the resulting Stokes waves are in phase and there is amplification by stimulated Raman scattering (Fig. 8.43c).

Stimulated Raman amplification is a nonlinear process, i.e., the power of the Raman wave (P_S) increases quadratically with the pump power (P_p) (see also Sect. 10.2.1).

For a simulation, one has to consider both the absorption of the pump wave in the z direction and the amplification of the signal wave in the z direction individually. If one pumps in forward *and* backward direction, one has to solve the following coupled differential equations:

- Amplification of the signal wave (Stokes) by Raman coefficient C_R and the summary power of the pump waves $P_P^-(z) + P_P^+(z)$ (+means forward direction, parallel to the signal direction, and—means backward direction, opposite to the signal direction) and the power of the signal wave $P_S(z)$, attenuation by the absorption with the absorption coefficient k_S

$$\frac{dP_S(z)}{dz} = -k_s \cdot P_S(z) + C_R \cdot \left[P_P^+(z) + P_P^-(z) \right] \cdot P_S(z)$$

$$(8.10)$$

- Attenuation of the pump waves by absorption with the absorption coefficient k_P, amplification of the pump waves by re-emission via the Raman coefficient C_R and by pump power $P_P^\pm(z)$ and signal power $P_S(z)$:

$$\frac{dP_P^+(z)}{dz} = -k_P \cdot P_P^+(z) + \left(\frac{\lambda_S}{\lambda_P}\right) \cdot C_R \cdot P_P^+(z) \cdot P_S(z)$$

$$\frac{dP_P^-(z)}{dz} = +k_P \cdot P_P^-(z-L) + \left(\frac{\lambda_S}{\lambda_P}\right) \cdot C_R \cdot P_P^-(z) \cdot P_S(z-L) \tag{8.11}$$

with initial values: $P_P^+(z=0) = P_0^+$ and $P_P^-(z=0) = P_0^- \cdot e^{-k_P \cdot L}$

In the small signal case (small values $\left(\frac{\lambda_S}{\lambda_P}\right) \cdot C_R \cdot P_S(z) << k_P$), **Eqs. 8.11** is simplified and the following equations are obtained

$$\frac{dP_P^+(z)}{dz} = -k_P \cdot P_P^+(z)$$

$$\frac{dP_P^-(z)}{dz} = +k_P \cdot P_P^-(z-L) \tag{8.12}$$

Solving Eqs. (8.12) results in an exponential decrease in pumping power along the fiber:

$$P_P^+(z) = P_0^+ \cdot e^{-k_P \cdot z} \text{ and } P_P^-(z) = P_0^- \cdot e^{+k_P \cdot (z-L)} \tag{8.13}$$

Substituting Eqs. (8.13) into (8.12) and integrating from 0 to L in a certain way solves Eq. (8.10) and obtains

$$P_S(z) = P_{S0} \cdot e^{-\frac{C_R}{k_P} \cdot [P_0^+ - P_0^- \cdot e^{-k_P \cdot L}]} \cdot e^{-k_S z} \cdot e^{-\frac{C_R}{k_P} \cdot [P_0^+ \cdot e^{-k_P \cdot z} - P_0^- \cdot e^{+k_P \cdot (z-L)}]} \tag{8.14}$$

You can normalize the power $P_S(z)$ to the value at $z=0$ and get $P_S^{norm}(z)$

$$P_S^{norm}(z) = e^{-k_S z} \cdot \frac{e^{-\frac{C_R}{k_P} \cdot [P_0^+ \cdot e^{-k_P \cdot z} - P_0^- \cdot e^{+k_P \cdot (z-L)}]}}{e^{-\frac{C_R}{k_P} \cdot [P_0^+ - P_0^- \cdot e^{-k_P \cdot L}]}} \tag{8.15}$$

From this one can calculate the gain G(z) in dB:

$$G(z) = 10 \, \log P_S^{norm}(z) \tag{8.16}$$

as well as the on-off gain $G_{\text{on-off}}(z)$ in dB

$$G_{on-off}(z) = 10 \, \log\left(\frac{P_S^{norm}(z)}{P_S^{norm}\left(bei \, P_0^\pm = 0\right)}\right) = 10\log\left(\frac{P_S^{norm}(z)}{e^{-k_S z}}\right) \tag{8.17}$$

Some results will be presented as examples (details can be found in the folder "Extras" as pdf file under "Raman Simulation").

For example, calculations were performed for a 50 km SSMF with parameters $C_R = 0.39$ W^{-1} km^{-1}; $k_S = 0.0463$ km^{-1}; $k_P = 0.053$ km^{-1} with forward (P_1) and reverse ($P_2 = P_1$) pumping. The pump wavelength was $\lambda_P = 1450$ nm, and the wavelength of the signal to be amplified was $\lambda_S = 1550$ nm. The calculations in MathCad can be traced in the folder "Extras" under "Raman SMF 50 km", of course there is also a corresponding pdf file. A similar calculation can be done for the dispersion compensating fiber DCF with 5.6 km length (MathCad or pdf file). As solution of the coupled differential equation both fibers are calculated in MathCad (SMF and DCF).

A major problem, which has not yet been adequately solved, arises from the fact that any nonlinear process (such as stimulated Raman scattering) becomes effective only at sufficiently high pump power. Another problem is the realizable gain. If one compares the gain in the EDFA (about 20–25 dB, see Fig. 8.34) with that in the Raman amplifier (about 8 dB at 200 mW pump power, see Fig. 8.44 or 8.46), the problem becomes clear - the 200 mW pump power used is already close to the distortion limit for single-mode fibers (SMF). With multiple pump wavelengths (see Fig. 8.47), the total pump power at one point is critical. For this reason, too, Raman amplifiers pumped at both ends are advantageous.

Exercise 8.8

What is the difference between spontaneous and induced Raman amplification? Which amplification is used in the Raman amplifiers?

▶   **Tip**
Help H8-8 (Sect. 12.1).
Solution S8-8 (Sect. 12.2).

8.4.2.4 Experimental Realization
For typical pump wavelengths, the glass fiber (single-mode fiber) is nearly transparent, i.e., pump light is absorbed only in long fibers.

One big advantage of Raman optical amplifiers is the possibility to "distribute" the amplification over a long distance. Typically, pumping from the backward direction is used (Fig. 8.49), and the components used (optical isolator, coupler, and bandpass filter) are the same as in EDFA.

However, it is also possible to operate the Raman amplifier in forward direction only or in forward *and* reverse direction. A corresponding setup from our laboratory is sketched in Fig. 8.50.

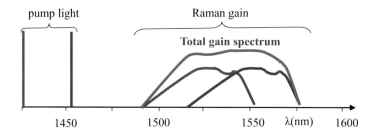

Fig. 8.47 Raman gain spectrum at synchronous pumping with 1425 *and* 1455 nm

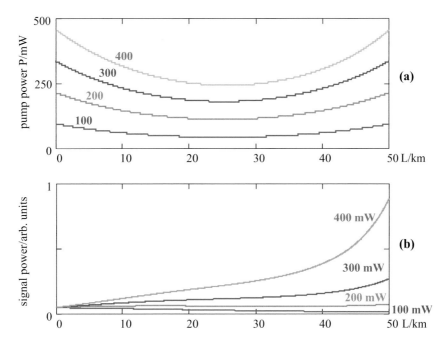

Fig. 8.48   Absorption (a) and Raman gain (b) in a 50 km SSMF at forward *and* backward pumps

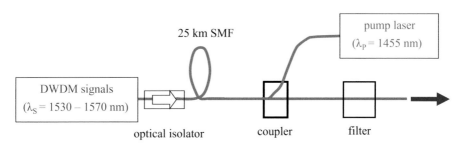

Fig. 8.49   Experimental setup of a Raman amplifier

Electrical signals (D_1 to D_n) are brought to a multiplexer (MUX) via transmitters (T_1 to T_n with different wavelengths λ_1 to λ_n and different powers P_1 to P_n). The light passes through optical isolators (OI), single-mode fibers (SSMF), and dispersion compensating fibers (DCF) to a demultiplexer (DeMUX) and is converted back to electrical signals (D_1 to D_n) in receivers (R_1 to R_n). The Raman active DCF is pumped in forward (pump with λ_{p1} and power P_P^+) or reverse direction (pump with λ_{p1} and power P_P^-) (Fig. 8.50). The whole is operated bidirectionally. In such a setup, in addition to dispersion compensation (re-shaping) according to Chap. 0), it was possible to achieve a Raman gain of up to 12 dB (or 16 times power gain).

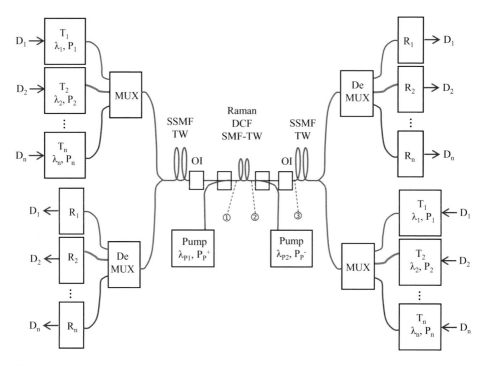

Fig. 8.50 Raman amplifier, pumped in forward and backward directions

---

Exercise 8.9

What are the advantages and disadvantages of the ROA compared to the EDFA?

▶ **Tip**
    Help H8-9 (Sect. 12.1).
    Solution S8-9 (Sect. 12.2).

8.4.2.5 Problems in Raman Amplifiers

For optimum amplification, the pump laser should have exactly the same polarization as the signal light to be amplified - any deviation results in a decrease in gain. This problem can be solved using two pump lasers with crossed polarization - but one has to pay the price of additional losses by coupling elements.

A problem arises if losses in the fiber (in the Third optical window, this is about 0.2 dB/km, thus we have about 5 dB in 25 km SMF) are higher than the generally positive) gain by "distributed" amplification along the fiber (Fig. 8.51). For example, we get a maximum amplification of 4 dB at 100 mW pump power; in this case the total amplification is negative - instead of an *increase,* we get a *decrease* in power (Fig. 8.51).

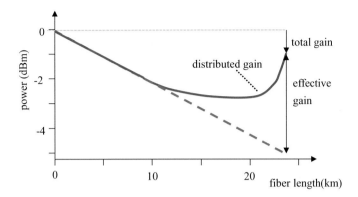

Fig. 8.51 "Distributed", effective and total gain in a Raman amplifier

We can solve this problem (Fig. 8.37) either by a reduced fiber length, or by an increased pump power. However, increased pump power can also result in the appearance of (disturbing) nonlinear effects.

With lower gain in ROA, we have to decrease the distance to the Raman amplifier which follows (shorter distance between amplifiers than for EDFA).

8.4.2.6 Noise in Raman Amplifiers

Like in EDFA, the main source of noise in Raman amplifiers is amplified spontaneous emission (ASE) of light at the wavelength of the signal. Due to "distributed" amplification, noise in ROA is less than in EDFA. Nevertheless, the large bandwidth of noise in the 6 THz range results in more noise. Unlike EDFA (see Fig. 8.37), the presence or absence of signal light does not play an important role in Raman optical amplifiers (ROA).

8.4.3   Semiconductor Optical Amplifiers (SOA)

Like in semiconductor lasers, in a **s**emiconductor **o**ptical **a**mplifier (SOA) we have stimulated emission of light. In Fig. 8.52 a double-hetero diode (DH) is depicted. In contrast to a DH-Fabry-Perot semiconductor laser, the end faces do **not** act as mirrors (e.g., by an anti-reflection coating). Figure 8.53 permits a look to the inner part of a semiconductor amplifier. The semiconductor amplifier has dimensions less than 1 mm (circle in Fig. 8.53).

Thus, we can use this Fabry-Perot arrangement *below laser threshold*. The amplification bandwidth is determined by resonances of the Fabry-Perot arrangement. For a "classical" Fabry-Perot semiconductor laser (semiconductor *with* end face reflection) operating below threshold, we can expect amplification bandwidth of some 10 GHz (up to about 0.1 nm). With anti-reflection coating one can get $\Delta f_v = 10$ THz (about 65 nm).

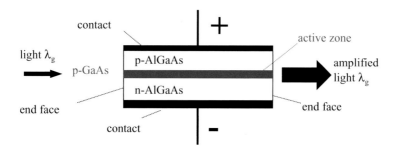

Fig. 8.52 Amplification in a semiconductor optical amplifier (SOA)

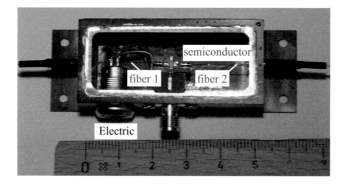

Fig. 8.53 Example of a SOA

Thus, the amplification bandwidth is like the emission bandwidth of an LED. A basic criterion we assume is an amplification change of maximum 1 dB; then we get $g \cdot R < 0.05$; g is the gain in dB; R is the "rest" reflectivity (for simplicity we take the same reflectivity for both end faces). With $R = 0.05\%$ or 0.005% one gets a maximum gain of $g = 20$ dB or $g = 30$ dB, respectively.

Exercise 8.10

What is the difference between an SOA and a Fabry-Perot laser?

▶   Solution S8-10 (Sect. 12.2).

Exercise 8.11

Why should the end faces of an SOA be anti-reflection coated?

▶   Solution S8-11 (Sect. 12.2).

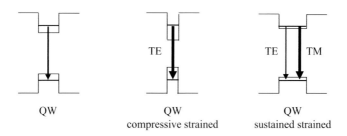

Fig. 8.54 "Normal", compressive and sustained strained quantum wall (QW)

Today, instead of volume semiconductors (so-called 3D semiconductors), quantum walls (quantum films, so-called 2D semiconductors) of several nanometres thickness (Fig. 8.54) are commonly used. At light amplification in MQWs, polarization plays an important role; in this case we have to distinguish between *transversal electric* (TE) and *transversal magnetic* waves (TM). In 3D semiconductors, amplification is equal for TE- and TM-waves; but in quantum walls, it is different.

To compensate the different amplification of TE- and TM-waves, one can use compressive *and* sustained strained quantum walls (QW). In compressive strained QWs, mainly TE-waves will be amplified; in sustained strained QWs, the TM-waves are amplified. Combining three compressive and three sustained strained QWs we get the spectral amplification depicted in Fig. 8.56 (after [Hul 96]. For both TE and TM polarization, we see the same amplification. The amplification bandwidth can be estimated from Fig. 4.20 as $\Delta\lambda_v = 55$ nm.

Details of the amplification spectrum of course depend on the operating current. One example for three different currents is depicted in Fig. 8.56. As one can see, the amplification spectrum for TE and TM polarization is different only for higher current (150 mA).

Noise in SOA again depends on amplified spontaneous emission (ASE). In SOA we have a noise factor of $F^* = 4 \ldots 7$ dB.

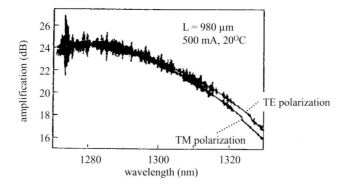

Fig. 8.55 Amplification spectrum of an MQW for two polarizations [Hul 96]

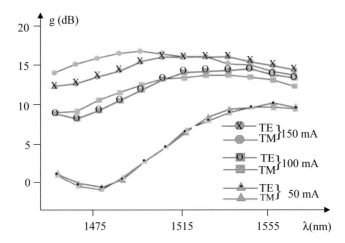

Fig. 8.56 Dependence of amplification spectrum on current (after [Hul 96])

8.5   Dispersion Compensation (Re-Shaping)

When light waves propagate in SMF, the bits widen due to dispersion, since the group refractive index ng(λ) is smaller for the "blue" component of the bit (component with lower wavelength) than for the "red" (component with higher wavelength), see also Sect. 3.4.1. As a result, "blue" component of the bit propagates faster than the "red"—this is the essence of dispersion. During re-shaping this process is reversed.

8.5.1   Application of Dispersion Compensating Fibers

In Chap. 3 it has already been described that the dispersion properties can be changed by clever selection of the refractive index profile or the ratio of core to cladding refractive index. As an example, three single-mode optical fiber types (SSMF, DSF as SMF with reduced core diameter, and DCF) with their most important parameters (core diameter d, mode field diameter $2w_0$, effective core cross section A_{eff} and chromatic dispersion parameter D) will be compared. The parameters used are shown in Table 8.3. The dispersion broadening and dispersion curve can be seen in Fig. 8.57. From Fig. 8.57 it can be seen that the zero dispersion in the DCF is around 1500 nm. However, it is unlikely that SSMF will be replaced by DSF on a large scale.

Comparing the chromatic dispersion of a standard fiber with the dispersion of a DCF (Fig. 8.57a and c), we see that the combination of 8 km of SMF (with D = +16 ps/km·nm) with 1 km of DCF (with D = −120 ps/km·nm) results in zero dispersion at $\lambda = 1500$ nm.

Table 8.1   Rare earth in fiber-optical amplifiers with amplification range

Rare earth	Formulae	Amplification range
Holmium	Ho^{3+}	1165–1205 nm
Präsodymium	Pr^{3+}	1270–1320 nm
Neodymium	Nd^{3+}	1315–1360 nm
Thulium	Tm^{3+}	1425–1475 nm
Thulium	Tm^{3+}	1605–1685 nm

Table 8.2   Parameters of various glass fibers

Fiber type	$C_{R,max}$ $W^{-1}\,km^{-1}$	α_P at 1450 nm dB/km	α_S at 1550 nm dB/km	D_{1550nm} ps/(km·nm)	S_{1550nm} ps/(km·nm2)
SSMF	0.39	0.23	0.19	16.5	0.058
TW-RS	0.71	0.25	0.20	4.5	0.045
TW-REACH	0.63	0.25	0.20	7.1	0.042
DCF für SSMF	2.45	0.54	0.41	−120	−0.422
DCF für TW-RS	3.50	0.75	0.58	−160	−1.600

Table 8.3   Selected parameters of SSMF, DSF, and DCF

	d (µm)	$2w_0$ (µm)	A_{eff} (µm²)	D_{Chrom} (ps/km·nm) at 1500 nm
SSMF	9	9,5–11,5	80	+16
DSF	5	8,6	58	+1,85
DCF		≈2,5	15–25	−120

With the dispersion values available, it is thus possible, for example, to compensate for the dispersion in 40 km of SMF by 5 km of DCF at 1550 nm (Fig. 8.58). The DCF is not laid as an optical fiber but is spliced in between two SMF in coiled form.

A disadvantage of such a setup is the increased attenuation in the DCF of about 0.5 dB/km compared to 0.2 dB/km in SMF—for the arrangement shown in the figure with two 5 km dispersion compensating fibers DCF one thus obtains 3 dB additional attenuation. In addition, the polarization mode dispersion increases. It is also important to consider the coupling between SSMF and DCF (see Sect. 4.1.2). From the formula mentioned there for different mode field diameters $a_w = -20 \cdot lg\left(\frac{2w_{0,SSMF} \cdot 2w_{0,DCF}}{w_{0,SSMF}^2 + w_{0,DCF}^2}\right)$ these losses result as about 0.9 dB.

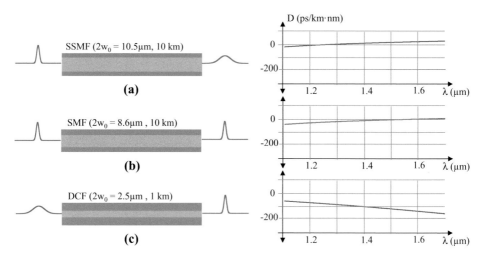

Fig. 8.57   Dispersion broadening and dispersion curve in SSMF (a), DSF (b), and DCF (c)

Exercise 8.12

Which basic idea plays the decisive role in dispersion compensation by DCF?

▶   **Tip**
Help H8-12 (Sect. 12.1).
Solution S8-12 (Sect. 12.2).

Exact compensation is of course only possible at one wavelength (here at about 1550 nm)—but by clever choice of the slope of the dispersion behavior one can achieve that the dispersion in the range of the Third optical window is less than $2\frac{ps}{nm\cdot km}$. In the practical experiment, a 10 Gbps signal was transmitted over a 617 km standard single-mode fiber, with a DCF path added every 50 km for dispersion compensation. It is also interesting to note that the order of SSMF and DCF is arbitrary (i.e., first 8 km DCF, then 40 km SSMF) and the lengths are scalable (i.e., first 120 km SSMF, then 24 km DCF).

Of course, there are many commercial offers for this today, e.g., from m2optics.

Fig. 8.58   Example of
dispersion compensation

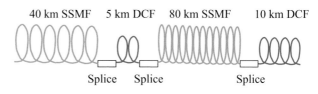

Exercise 8.13

The dispersion in a SMF ($D_{\mathrm{chrom}} = 18\frac{\mathrm{ps}}{\mathrm{nm \cdot km}}$) should be compensated with a DCF ($D_{\mathrm{DCF}} = -100\frac{\mathrm{ps}}{\mathrm{nm \cdot km}}$). What length of SMF can be compensated with 10 km DCF?

▶ **Tip**
   Help H8-13 (Sect. 12.1).
   Solution S8-13 (Sect. 12.2).

8.5.2   Dispersion Compensation with Fiber-Bragg-Gratings

A bit consists of different spectral components, in Fig. 8.59 three wavelengths λ_{-1}, λ_0, and λ_{+1}, as well as the envelope are shown as an example at the fiber input. The left region with λ_{-1} is also called the "red" part of the bit and the right one with λ_{+1} the "blue" part. At the fiber input, all wavelengths are represented at the same time—you can think of it as the sum of Gaussian pulses with different wavelengths. After passing through an optical fiber of length L, dispersion occurs and the bit at the fiber exit in Fig. 8.59 is broadened. The reason for this is the different propagation velocity (group velocity $v_g(\lambda) = \frac{c_0}{n_g(\lambda)}$) with n_g as group index) of the "red" and the "blue" part of the bit. According to Fig. 3.16, the group index n_g is larger for the "red" than for the "blue" part of the bit, and the group velocity vg is smaller for the "red" than for the "blue" part of the bit. As a result, the "blue" part of the bit arrives earlier than the "red" part at the coupler and then at the fiber Bragg grating (FBG), similar to the one shown in Fig. 8.20.

In contrast to Fig. 8.20, however, the lattice constants of different parts of the FBG are now different - near the FBG input the lattice constant is adapted to the "red" part of the spectrum, afterwards a lattice constant is adapted to the "green" and at the end to the "blue" part of the spectrum. Thus, first the "red" part of the bit is reflected at the

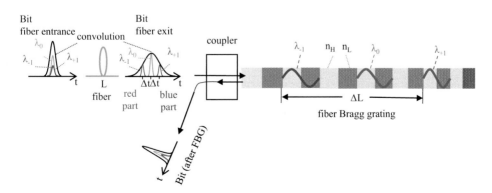

Fig. 8.59 Application of a fiber Bragg grating for dispersion compensation

FBG and coupled out via the coupler as a "bit after the FBG" - "green" and "blue" parts follow with a time delay. As a result, if the distance between the gratings ΔL is chosen appropriately, the dispersion is compensated, and the bit is restored.

In the example we have used only 3 wavelengths, in reality there are infinitely many. This can be considered by using many different grating constants of the FBG - or by changing the grating constant continuously and obtaining a so-called "chirped" grating. Decisive is the transit time difference for outgoing and returning wave $2\Delta L/v_g = 2 \, ng \cdot \Delta L/c_0$. If this agrees with $2\Delta t$ then one obtains complete dispersion compensation. Full matching is not necessary - over- or under-compensation of chromatic dispersion is obtained.

Dispersion compensation with FBGs is scalable. For this purpose, one FBG with a range of matched grating constants must be installed at different locations on the optical fiber for each channel (wavelength). However, one has to consider that due to the different propagation times the dispersion compensated channels occur staggered in time.

Exercise 8.14

What is the basic idea for dispersion compensation with FBGs?

▶   **Tip**
     Help H8-14 (Sect. 12.1).
     Solution S8-14 (Sect. 12.2).

8.6   Clock Recovery (Re-Timing)

In serial communication of digital data, re-timing (clock recovery) is the process of extracting timing information from a serial data stream itself, allowing the timing of the data to be accurately determined without separate clock information. Essentially, it involves an unwanted shift in phase. Causes are (Fig. 8.60):

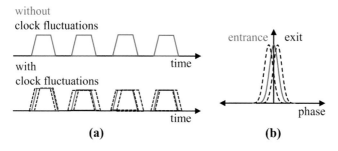

Fig. 8.60   On the origin of clock fluctuations due to jitter (a) and phase fluctuations (b)

- Jitter (fluctuation) in the AC range (e.g., fluctuation of the clock frequency, fluctuations in the time domain).
- Phase offset in the DC range (e.g., random difference of the phase of the output pulse to the phase of the input pulse).

When the bits are spread over a long distance, a "smearing" of the bits in time and/or in phase occurs and thus a reduction of the bit rate (BR), which in turn results in the necessity to increase the signal-to-noise ratio (SNR) (see also Sect. 7.4.5, Fig. 7.11).

The basic idea is therefore to recover the clock (bit rate) from the signal itself and thus to undo the "smearing". Two different methods are used for clock recovery:

- Nonlinear methods of input signal processing.
- Decision feedback methods with adaptive filters.

Both methods work with a correlation of the incident pulses.

Exercise 8.15

What is the basic idea behind clock recovery?

▶ **Tip**
Help H8-15 (Sect. 12.1).
Solution S8-15 (Sect. 12.2).

8.6.1 Clock Recovery by Using Nonlinear Methods

In this method, the signal edges are obtained by nonlinear methods of processing the input signal, which then result in the symbol clock with a band pass and subsequent threshold decision. In principle, various nonlinear functions can be used (see Chap. 10), such as squaring (generation of the harmonics) or differentiation (parametric effects) with subsequent magnitude formation (rectification) of the input signal.

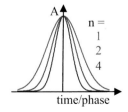

Superposition (formation of the correlation function) reduces the random "smearing" of the pulse (bit). The superposition can be generated by 2 or multiples of 2 from Gaussian pulses:

$$A(t) = \left(A_0 e^{-x^2}\right)^n \text{ with } n = 1, 2, 4 \ldots$$

The result can be seen in the adjacent figure. From the shortened pulses the approximate time or phase position and thus the clock is determined.

The disadvantage of this method is that the exact phase position cannot be recovered with it. In practical systems, the exact sampling time of the received data is then usually made by a phase offset (time offset) that is fixed in the receiver. This time offset represents the exact sampling time within the duration of a symbol. This time offset must therefore be derived from the respective frequency in multi-frequency systems.

8.6.2   Clock Recovery by Decision Feedback

This clock recovery method makes use of the so-called first Nyquist condition. This condition states that the temporal symbol crosstalk, also called intersymbol interference, disappears. The prerequisite for this is the use of adaptive channel equalizers for compensation in the case of practically all transmission channels and their distortions.

If the transmission channel is appropriately equalized and satisfies intersymbol freedom, the transmit clock can be recovered exactly by synchronizing to the zero crossing of the receive signal before and after the individual transmit pulses. This information is time-averaged and serves as a control variable for a controllable oscillator.

The advantage of this method is that it not only accurately recovers the sampling frequency, but also the sampling phase. A disadvantage is the effort associated with adaptive filters. Above all, the necessary training sequence at the beginning of a transmission must be provided for in the transmission protocol.

One form of clock extraction using a feedback control loop is the application of a modified Costas Loop (Costas Loop). The Costas loop was developed in 1956 by John Peter Costas (1923–2008). Usually, this loop is used for coherent demodulation of digital phase modulations (e.g., phase-shift keying). Instead of synchronizing to the carrier frequency, an appropriately modified Costas Loop can also be used to recover the symbol clock. In this case, the phase position is also recovered correctly.

The method has not yet been used commercially. As an example, a setup will be explained that was developed at the University of Strathclyde, Glasgow (UK) [Gle 13] and is particularly suitable for compensation of interference between many users (multiple access interference, MAI) (Fig. 8.61a). Data are transmitted bidirectionally at a bit rate of 2.5 Gbps (corresponding to a bit spacing of 400 ps with a bit duration of 8 ps) over a 17 km long optical fiber with chromatic dispersion compensation. Erbium-doped

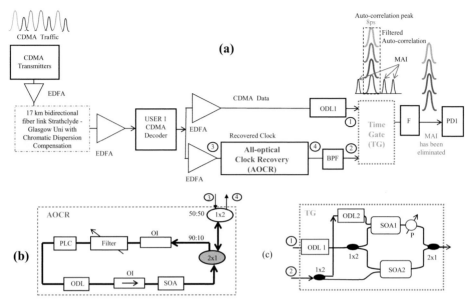

Legends: MAI – Multiple-User Access Interference, EDFA – Erbium Doped Fiber Amplifier, SOA – Semiconductor Optical Amplifier, PD – Photo Detector, ODL – Optical delay Line, BPF – Band Pass Filter, OI – Optical Isolator, PLC – Polarization Controller, P - Phase adjustment heating element, TG – Time Gate, AOCR – All-Optical Clock Receiver

Fig. 8.61 Clock Recovery: (a)—total arrangement, (b)—clock extraction, and (c)—gate structure (after [Gle 13]

fiber amplifiers (EDFA) are used as amplifiers. The signal at user 1 is decoded and then split with a 3 dB coupler into a line to the data line and a line to the clock recovery.

Figure 8.61b describes clock recovery using a modified Costas loop (all-optical clock recovery, AOCR). The signal passes through a coupler and is fed into the loop via a 2×1 coupler. After passing through the elements shown in Fig. 8.61b (including polarization control PLC and semiconductor optical amplifier SOA), the signal comes back to the 2×1 coupler. 10% is decoupled there. 90% is sent back to the loop, overlapped with the signal of the first pass (autocorrelation), and so on. The correlation amplifies the signal in the clock and thus the clock is recovered.

The recovered clock is now merged with the data line via an optical time gate (Time Gate TG in Fig. 8.61c). Via a filter, the signal with eliminated MAI is thus available at the receiver PD1.

The procedure seems very complex and not very suitable for widespread use—but it solves the problem of clock recovery very well.

Measurement Technology in Optical Fibers and Optical Transmission Systems

9

Today it is becoming increasingly important to check the error-free and high-quality operation of optical networks—especially at high bit rates—on a regular basis. This means that one has to check parameters of all optical elements. Because glass fibers are the most important element in a fiber-optical system, the main attention should be given to fiber parameters. Note that it is important to inspect these parameters *during* network operation. Now we will discuss the main aspects of measuring technology with the different measuring techniques.

9.1   Measurement Technology in Glass Fibers

In this chapter, we will describe several - from the viewpoint of the author - important measuring methods for fibers. Further detailed information is published elsewhere [Brü 98].

9.1.1   Measurement of Index Profile

Besides the measurements of fiber geometry (core and cladding diameters, flatness of front surfaces), measurements of the refractive index *along the whole fiber* are of importance. However, all common measuring methods are performed only at a certain position - i.e., mostly at the front surfaces.

Supplementary Information The online version contains supplementary material available at
https://doi.org/10.1007/978-3-658-43242-3_9.

9.1.1.1 *Near-Field Scanning Method (Near-Field Scanning)*

The so-called near-field of light is the distribution of power density P at the end surface of glass fiber. Because in multi-mode fibers the power distribution is proportional
to the distribution of refractive index, we can perform measurements of power distribution to get the refractive index. Light is guided in a glass fiber of suitable length (less
than 2 m) to obtain a homogeneous distribution of light between all modes (*equilibrium
mode distribution*, see Sect. 3.2.3). By one or several lenses we will enlarge the fiber end
surface to a screen. Now one has to analyze the intensity of the whole picture (e.g., by
a CCD camera) or pointwise by moving the diaphragm and the receiver D together with
the lens (Fig. 9.1a). If we are certain that we have a symmetric refractive index profile,
it is enough to move the receiver only in one direction—otherwise one has to measure in
two directions orthogonal to each other. Mostly, measurements today are computer controlled and we get, e.g., results as depicted in Fig. 9.1b. The solid line is characteristic for
GI fibers, dotted line for SI fibers. In addition, near-field scanning also gives information
on core diameter d.

9.1.1.2 *Refracted Near-Field Method*

Requirements with regard to minimum fiber length (to obtain a homogeneous light distribution between modes) can be avoided using the refractive near-field technique. Light
is focused by a microscope object lens to a point (spot) of the entrance surface (Fig. 9.2).
In this method the angle of beam spread (2Θ) should be larger than the doubled angle of
acceptance ($2\varphi_A$) - thus we have no problems with disturbing light in the core. After a
certain distance, this light will leave the fiber. The fiber is in a cell filled with an immersion liquid of refractive index similar to glass. We measure the intensity of light *not
distributed* into the fiber (Fig. 9.2) - the "leftover" (which is the important part for communication) will be blocked by an opaque aperture. If one moves the position of the spot
(r) with respect to the center of the fiber, the local refractive index n(r) is changed and
therefore the exit angle. This then results in changes in the measured light power P(r).
From measurements of P(r) one can calculate the refractive index profile.

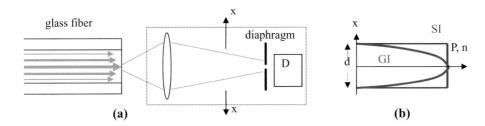

Fig. 9.1   Near field scanning (a) and examples (b)

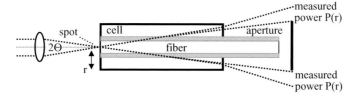

Fig. 9.2 Refracted near-field technique

9.1.2 Measurements of Losses

Attenuation of light and the resulting *attenuation coefficient* (attenuation per kilometer fiber length) are important parameters in wave guide technology. There is an additional problem for multi-mode fibers. Due to the existence of *transversal modes*, the light distribution between modes is changed at the beginning of the fiber, but a stable distribution is a basic principle of each exact measurement. Thus, one has to use suitable procedures to reach a stable distribution, e.g., using *mode mixers* or *mode strippers*, see Sect. 3.2.3). The *influence of measuring conditions* is of high importance for exact attenuation measurements and should be kept in mind when considering the descriptions of measuring methods which follow. Otherwise, the values of attenuation measured could be too high.

9.1.2.1 Ut-Back and Substitution Methods

Measurement of attenuation by cut-back and substitution methods is sketched in Fig. 9.3. Light at a certain (in general changeable) wavelength λ of power p_0 will be used. To obtain a stable mode distribution, the procedures described above must first be performed - the simplest way is mode mixing by a 100 m launching fiber. Then, the light is coupled by a coupler into the long fiber which is being analyzed.—the power at the entrance of this fiber is p_0' and cannot be measured. At the end of the fiber of length L_1, we can measure the power p_1 directly by a receiver. The attenuation can be determined as the difference $a_1 = p_0' - p_1$. To eliminate the (still unknown) power p_0' one has to perform a second measurement with a short ($L_0 \approx 1$ m) fiber truncated from the long fiber (cut-back method) and we measure the power p_2. The attenuation is $a_2 = p_0' - p_2$. Thus, we get the attenuation of the long fiber $L = L_1 - L_0$ as a difference of two attenuation

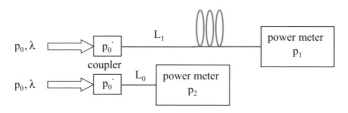

Fig. 9.3 Cut-off method

measurements $a=a_1 - a_2 = p_2 - p_1$, attenuation coefficient is $\alpha=a/L$. The substitution method is based on the same idea - instead of a cut-off, one uses a short fiber of the same material and with the same properties as the fiber to be measured.

Exercise 9.1

Light at 1300 nm wavelength with power $p_0=0$ dBm passes through a 10 km fiber, after that $p_1=-7$ dBm is measured. After a short (1 m) fiber $p_2=-2$ dBm is measured. How much is the attenuation coefficient?

▶ **Tip**
    Help H9-1 (Sect. 12.1).
    Solution S9-1 (Sect. 12.2).

If one repeats these measurements for different wavelengths (e.g., in the range of optical communication between 750 and 1600 nm), one gets the *attenuation coefficient versus wavelength* of fibers. Important for good quality measurements is that coupling between launching and long or short fiber by plugs is not changed during measurements (cut-off method) or should be identical (substitution method).

9.1.2.2 Backscattering, OTDR Method

Optical Time Domain Reflectivity (OTDR) is based on backscattering of light (Fig. 9.4). A light pulse of certain duration τ_p (between 100 ps and 1 µs) and wavelength (mostly wavelength of an optical window—850 nm, 1300 nm or 1550 nm) will be coupled into the fiber being studied. At the same time the pulse "starts" a "clock" in the oscilloscope. With this scope we measure the time of flight of the pulse to the so-called incidence (e.g., a slice, a plug, a fiber break, or the end of the fiber) and back via coupler to the receiver D—and therefore to the scope. The amplitude and shape of backscattered incidence signal of power p_R contain information (e.g., about fiber break). If the group index n_g is known, one can calculate from time t_E of the incidence, its position L_E using the following formula:

Fig. 9.4 Basic principle of OTDR measurements

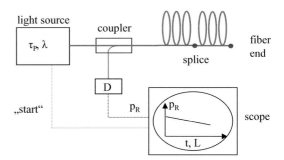

$$L_E = \frac{t_E \cdot c}{2n_g} \tag{9.1}$$

with c - speed of light. If we have for example a fiber break ($n_g = 1.5$) after 25 µs, the fiber is broken at $L_E = 2.5$ km. Other incidences like splices or plugs result in an additional attenuation.

If the group index n_g is known, we automatically get the distance L of the incidence in standard commercially available OTDR equipment.

By means of some real OTDR measurements performed in our lab (Fig. 9.5) we would like to discuss some typical problems. The measurements were performed in the first optical window at 850 nm using multi-mode fibers (MMF). The group index (e.g., $n_g = 1.4860$) as well as the used pulse duration (τ_P) are known or can be adjusted. Furthermore, one can select the measurement range of distance and power.

If one considers the measurement (Fig. 9.5) first we note that there is a strong attenuation in the first 100 m part of the fiber - this is the influence of measuring conditions (redistribution of energy between modes, adjustment of mode equilibrium). The straight line after this permits the identification of the attenuation constant of the fiber ($\alpha = \Delta p / \Delta L = 2.6$ dB/km).

After the splice we find the same fiber with the same straight line as before. Then two identical fibers are connected by a plug. At 3.3 km distance, the fiber is broken—or, we have the end of the fiber (as one can interpret the high peak followed by noise).

Inserts in Fig. 9.5 show a splice and a plug in greater detail. A plug is characterized by a characteristic peak resulting from the space between two connected fibers. In a splice, this peak is missing. The position (or distance from the beginning) is always taken at the very beginning of incidence (splice or plug) - here we find $L_{Splice} = 0.97$ km

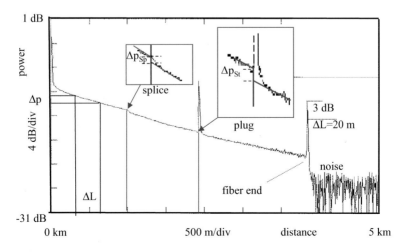

Fig. 9.5 OTDR measurement with splice and plug (pulse length $\tau_P = 100$ ns)

or $L_{Plug} = 1.92$ km, respectively. The attenuation connected with the incidence can be calculated in the following way: power at the beginning of incidence minus power (Δp_{Splice} or Δp_{Plug}), resulting from backward extension of the straight line of the fiber which follow (see inserts in Fig. 9.5). From inserts in Fig. 9.5 one can find attenuation of splice $\Delta p_{Splice} = 0.2$ dB and plug $\Delta p_{Plug} = 0.3$ dB, respectively.

Exercise 9.2

How can one distinguish between plug and splice in an OTDR measurement?

▷ **Tip**
Help H9-2 (Sect. 12.1).
Solution S9-2 (Sect. 12.2).

As one can see in Fig. 9.5 space resolution is determined by the "seeming length" ΔL of the splice (3 dB decay in Fig. 9.5) to about 20 m. Because the real "length" of the splice is only a few micrometers, the length ΔL is named the *dead zone*, i.e., one cannot distinguish whether we have one or more incidences. For example, several splices with a few meters distance would give a similar result. Length of the dead zone ΔL can be calculated from pulse duration τ_p by

$$\Delta L = \frac{\tau_p \cdot c}{n_g} \tag{9.2}$$

For measurement shown in Fig. 9.5 it gives $\Delta L = 20$ m.

Length of the dead zone can be decreased using shorter pulses (down to about 100 ps) - on the other hand laser energy is decreased, which results here in an increasing noise and therefore in a decreasing maximum distance to be measured. This is illustrated in Fig. 9.6.

This noise can be decreased, averaging over a large number of pulses (measurements). Averaging over 512 measurements is shown in Fig. 9.7.

Fig. 9.6 OTDR measurement with noise at $\tau_p = 3$ ns

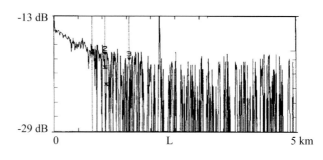

Fig. 9.7 OTDR measurement at $\tau_p = 3$ ns with averaging

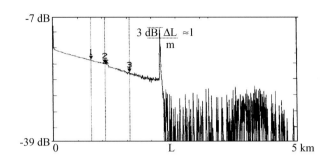

An OTDR measurement with $\tau_p = 20$ ns shows a splice with 1 dB attenuation. How can this result be interpreted?

▶   **Tip**
Help H9-3 (Sect. 12.1).
Solution S9-3 (Sect. 12.2).

A simplified version of OTDR is the pulse-echo method. In contrast to OTDR, at pulse-echo method, the receiver is much less sensitive (corresponding to a much lower price). That's why one can measure only fiber breaks and fiber ends (and therefore the length of fiber), but it is enough for practice. On the other hand, by using OTDR one is able to check fiber breaks, fiber ends, splices and plugs, and the corresponding attenuation.

9.1.3   Dispersion Measurement

If we have a sequence of several short (needle-shaped) light pulses with regular time intervals, at propagation in a fiber these pulses will be extended due to several mechanisms of *dispersion* (e.g., *modal, material, waveguide and/or chromatic dispersion*). This prolongation (the so-called *group delay time* Δt_g, see Sect. 3.4.1) results in a maximum fiber length; after that pulses are no longer distinguishable from each other. Thus, we get a minimum time interval where neighboring pulses are still distinguishable after passing through the fiber.

On the other hand, the light at the entrance of the fiber has a certain amplitude. If one modulates this amplitude with a modulation frequency f_M, the amplitude is decreased with increasing modulation frequency. This means glass fibers act as *low-pass filter*: at low modulation, frequencies signal passes through the fiber with low losses; at higher frequencies, the amplitude is reduced. Modulation frequency where the amplitude is decreased to 50% (or 3 dB) is the bandwidth B. There is a connection between bandwidth B and group delay time Δt_g, for Gaussian or bell-shaped pulses, we can use the formula:

$$\Delta t_g = \frac{0,44}{B} \tag{9.3}$$

Therefore, we have to measure only one parameter (Δt_g or B)—the other one can be calculated. It results in two different methods for dispersion measurement: *measurements in the time range* and *measurements in the frequency range*.

9.1.3.1 Measurements in Time Range

As described above, for dispersion measurements in time range we have to use a scope to measure the group delay time Δt_g at different distances of propagation in the fiber. Δt_g we can get as the difference between half-width (measured at 50% power decay) at the beginning (τ_p') and the end (τ_p'') of a fiber. Here we considered that pulses at the beginning already have a certain pulse length (Fig. 9.8) - thus we get $\Delta t_g = \sqrt{\tau_p''^2 - \tau_p'^2}$.

During propagation in a fiber, pulses will be also attenuated—thus in practice one measures the rise (increase of amplitude from 10 to 90% of maximum) and fall times (decrease of amplitude from 90 to 10% of maximum - one gets the pulse rise (t_{rice}) and pulse fall times (t_{fall}) at the beginning (t_{rice}', t_{fall}') and at the end of a fiber of length L (t_{rice}'', t_{fall}''). Group delay time Δt_g and dispersion parameter D can be calculated by

$$D = \frac{\Delta t_g}{L} = \frac{\sqrt{\left(\frac{t_{rice}''+t_{fall}''}{2}\right)^2 - \left(\frac{t_{rice}'+t_{fall}'}{2}\right)^2}}{L} \tag{9.4}$$

Theoretically the bandwidth B can be calculated by Eq. (9.3) as $B = \frac{0.44}{\Delta t_g}$.

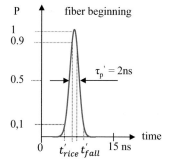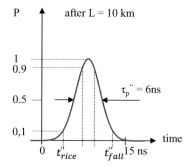

Fig. 9.8 Example of the group delay time of a bit in a glass fiber

Exercise 9.4

Which bandwidth can be expected in a 10 km glass fiber with dispersion parameter $D = 0.4 \frac{ns}{km}$?

▶ **Tip**
   Help H9-4 (Sect. 12.1).
   Solution S9-4 (Sect. 12.2).

Unfortunately, in practice one has much more complicated pulse shapes than are depicted in Fig. 9.8. For these cases, bandwidth can be found performing computer analysis using the so-called Fourier analysis.

9.1.3.2 Measurements in Frequency Range

By means of the so-called *wobbling method,* one can measure dispersion in the frequency range. Amplitude of a transmitter is modulated with a continuously increasing frequency f_M. By a receiver we measure the power. If we are able to keep the amplitude constant for all modulation frequencies (100% value) we get the transfer function directly; then the bandwidth is the frequency of the 50% or -3 dB value (e.g., 1 GHz in Fig. 9.9). Otherwise, if power is changed with modulation frequency f_M, we have to normalize the measured power, i.e., we have to take the difference between measured power and power at the transmitter side. From bandwidth B one can again calculate group delay time.

9.2   Measurement of Quality of Data Transfer

9.2.1   Measurements of the Bit-Error Ratio, Receiver Sensitivity

For optical data transmission it is very important to know which bit rates can be transmitted over which distance without correction of amplitude (attenuation) or of group delay time (dispersion). To this end we have to measure the Bit-Error Ratio (BER) as number of bit errors per number of transmitted bits - remember to not confuse this with the Bit-Error Rate, which is the number of bit errors per unit of time. The setup is depicted in Fig. 9.10. In a bit generator, a bit pattern is created at the corresponding

Fig. 9.9 Wobbling method

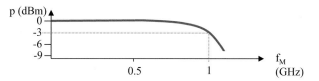

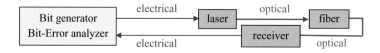

Fig. 9.10 BER setup

Fig. 9.11 Bit at high bit rate

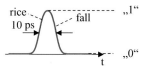

transfer rate. This bit pattern controls the laser. Thus, we get an optical bit pattern which propagates through the fiber to the receiver. By receiver the optical bit pattern is again converted to an electrical one. In a Bit-Error-Analyzer the number of bit errors is measured and compared with the total number of bits - we get the Bit-Error Ratio.

This setup is suitable for performing measurements in a glass fiber system - for permanent in situ measurements; however, it cannot be used because bits should be generated separately and transferred to the receiver. The sensitivity of the receiver is the minimum power at the receiver needed to realize a certain BER. Alternatively, a variable attenuator instead of the fiber can be used to measure the receiver sensitivity.

9.2.2 Eye Diagram

A low-priced alternative to the BER measurement of Fig. 9.10 is measurement of eye diagrams. This method can be used to check the quality of a *running optical transmission* on a permanent, long-term basis.

Eye diagrams work with digital signals, i.e., bits. As it was described in Sect. 6.1 at high bit rates in the range of 10—100 Gbit/s a single bit looks like a Gaussian pulse (Figure 9.11). In Figure 9.11 "1" denotes the logical binary 1, and "0" is the logical binary 0 for the rise and fall times.

During propagation in an optical transmission system, the bit experiences statistical influences which can change the pulse shape of Figure 9.11 significantly:

- Jitter of amplitude or power by noise (Fig. 9.12a); the lower the Signal-Noise Ratio (SNR, see Sect. 7.4), the higher the jitter in amplitude.
- Statistical smearing and bit broadening by dispersion (Fig. 9.12b).
- Statistical jitter in the clock cycle of bits resulting in a time jitter in pulse position in time (Fig. 9.12c).

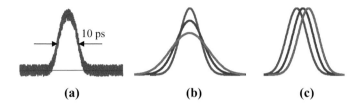

Fig. 9.12 Amplitude noise with SNR$=10$ (a), dispersion broadening (b), and jitter of clock cycle (c)

All these influences are especially strong at shot pulses, i.e., at extremely high transmission rates (e.g., 10 ps in Fig. 9.12).

For an eye diagram one uses, e.g., a bit sequence (see also corresponding MathCad program or pdf file) like in Fig. 9.13 with influence of noise and disturbances. All we have to know is the clock cycle (= bit rate). If we know the bit rate (BR) we also know the time distance Δt between two neighboring bits as $\Delta t = \frac{1}{BR}$. Then the pulse duration τ_P of single bits is either the reciprocal value of bit rate $\tau_P = \frac{1}{BR}$ at Non-Return-to-Zero (NRZ) operation or shorter than $\frac{1}{BR}$ at Return-to-Zero (RZ) operation. In general pulse takes half of the time available $\frac{1}{2 \cdot BR}$. In Fig. 9.13 a NRZ sequence of signals at $\tau_P = \Delta t$ is depicted.

By a sampling scope one measures the interference of all bits of pulse duration τ_P. Using the calculated bit sequence of Fig. 9.13 we get the eye diagram without or with white noise like shown in Fig. 9.14.

A more realistic picture of the eye diagram is given in Fig. 9.15. By noise and dispersion the primary perfect eye (Fig. 9.15a) is changed to a real one (Fig. 9.15b). For high-quality transmission the eye should be "open" - otherwise transmission is bad.

Fig. 9.13 Bit sequence 1–0 – 1 – 1 – 1 – 0 – 0 – 1 with frames of pulse duration τ_P

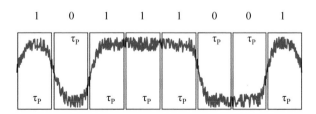

Fig. 9.14 Eye diagram without (a) and with (b) noise of the bit sequence of Fig. 9.13

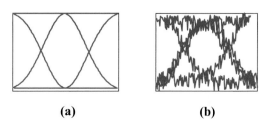

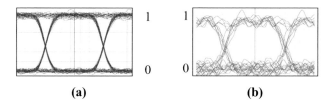

Fig. 9.15   Formation of eye diagrams: Nearly perfect (a) and eye diagram with noise (b)

Fig. 9.16   Example of an eye diagram

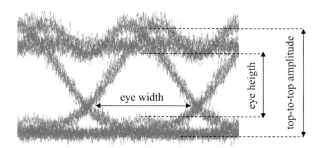

An example of a real eye diagram one can be seen in Fig. 9.16.

The most important parameters are indicated in Fig. 9.16. From eye height and width, one can calculate the amplitude and deformation of signals in terms of:

- Delay time resulting in quality changes in voice transmission.
- Bit-error rates; important for data and video transmission with high bit rates. For Ethernet transmission a BER of 10^{-8} is requested, for the token ring - $10^{-9,}$ and for Fiber Distributed Data Interface (FDDI) - $2.5 \cdot 10^{-12}$. For FDDI, this means that one (single) wrongly transmitted bit is allowed for a total of 400 billion bits transmitted.
- Noise: it results in a lower signal–noise ratio (SNR) and thus in an increase in bit-error rate.
- Jitter: i.e., phase fluctuations and therefore changes of signal frequency in time. Here we mean fluctuations of the (generally fixed) point in time when one amplitude changes to another digital signal amplitude (clock). We can expect this jitter especially at high frequencies; it can result in data losses.

Exercise 9.5

What does "the eye is nearly closed" mean for the quality of data transfer?

▶   **Tip**
Help H9-5 (Sect. 12.1)
Solution S9-5 (Sect. 12.2)

Nonlinearities in Glass Fibers

10

In today's technology where horizons are constantly expanding and new limits continually being achieved, nonlinearities and nonlinear behavior play a major role. Nonlinearities are ubiquitous in many processes, but we only become aware of this if a certain parameter (in optics the electric field strength E) is enlarged.

Much can be deduced from the consideration of classical mechanics. Newton's law states that the repulsive force (i.e., mass m multiplied by the acceleration $\frac{d^2x}{dt^2}$) is proportional to the displacement x according to Hook's law. The constant is the spring force k. Nonlinear behavior occurs when terms occur in addition to the linear deflection, i.e., terms with x^2, x^3, etc. (10.1).

$$m \cdot \frac{d^2x}{dt^2} = k \cdot x + k_1 x^2 + k_2 x^3 + \cdots \tag{10.1}$$

As a solution of the linear equation ($k_1 = k_2 = \ldots = 0$), consider the known cosine oscillations of the spring oscillator with angular frequency $\omega = 2\pi f = \sqrt{\frac{k}{m}}$.

$$x(t) = x_0 \cos \omega t \quad \text{or} \quad x(t) = \frac{x_0}{2} \cdot \left(e^{-i\omega t} + e^{i\omega t}\right) \tag{10.2}$$

At high amplitude, Hook's law becomes nonlinear - at high displacement one gets harmonics (vibrations at doubled frequency, see solution of Eq. (10.1)). Radio engineers know that at high transmitter power, we get harmonics at 2f, 3f, etc. We have already discussed laser characteristics as an example of nonlinear behavior: Above certain pump intensity, spontaneous emission of light becomes stimulated, and its intensity rises dramatically (see Sect. 5.4.1). Thus, we get the nonlinear P-I characteristic (Fig. 5.20).

Supplementary Information The online version contains supplementary material available at https://doi.org/10.1007/978-3-658-43242-3_10.

Nonlinearities in optics arise whenever the dielectricity or susceptibility ε or the refractive index n can no longer be considered as a constant but depends on another quantity (in optics the field strength E or the peak power for pulses). For example, *nonlinear optics* when using laser light with extremely high peak powers is based on these nonlinearities. The treatment according to the beam model is impossible because all waves must be considered as independent of each other. Therefore, a mathematical consideration is always linked to the wave Eq. (2.7):

$$\Delta \vec{E}\left(\vec{r},t\right) - \frac{n_0^2}{c^2}\frac{\partial^2 \vec{E}\left(\vec{r},t\right)}{\partial t^2} = 0 \, (linear \ case)$$

where n_0 is the refractive index at low field strength (up to now the "constant" refractive index).

In the nonlinear case, instead of zero, one has to consider contributions of nonlinearity on the right side of the equation, the so-called *nonlinear driver terms*.

$$\Delta \vec{E}\left(\vec{r},t\right) - \frac{n_0^2}{c^2} \cdot \frac{\partial^2 \vec{E}\left(\vec{r},t\right)}{\partial t^2} = \frac{1}{2} \cdot \frac{\partial^2 \left(\chi^{(2)}\vec{E}\cdot\vec{E}\right)}{\partial t^2} + \frac{1}{c^2} \cdot \frac{\partial^2 \left(\chi^{(3)}\vec{E}\cdot\vec{E}\cdot\vec{E}\right)}{\partial t^2} + \dots (nonlinear \ case)$$

(10.3)

The term $\chi^{(2)}\vec{E}\cdot\vec{E}$ is considered as a second-order nonlinearity, the term $\chi^{(3)}\vec{E}\cdot\vec{E}\cdot\vec{E}$ as third-order nonlinearity, etc., where $\chi^{(2)}$, $\chi^{(3)}$, etc. are the nonlinear coefficients, respectively. Second-order nonlinearities play a role only in crystalline materials, so they can be neglected for glass fibers. In contrast, third-order nonlinearities always play a role when dealing with unstructured materials such as glass.

In the nonlinear case, one always has to deal with several fields which add up to the total field strength E_{total}. For example, 4 fields contribute to the total field strength E_{total} in the case of third-order nonlinearities:

$$\vec{E}_{total} = \sum_{j=1}^{4} \vec{E}_j$$

$$with \quad \vec{E}_j = \frac{1}{2}\left[\vec{E}_{j0}\left(\vec{r},t\right)\cdot e^{-\left(\omega_j t - k_j z\right)} + c.c.\right]$$

Thus, the wave Eq. (10.3) must be considered separately for all 4 coupled waves:

$$\left(\nabla^2 + k_1^2\right)E_{01}e^{-i[\omega_1 t - k_1 z]} = -\frac{\omega_1^2}{c^2}\chi^{(2)}E_{03}\cdot E_{02}^*\cdot E_{04}^* e^{-i[\omega_1 t - (k_3 - k_2)z]}$$

$$\left(\nabla^2 + k_2^2\right)E_{02}e^{-i[\omega_2 t - k_2 z]} = -\frac{\omega_2^2}{c^2}\chi^{(2)}E_{03}\cdot E_{01}^*\cdot E_{04}^* e^{-i[\omega_2 t - (k_3 - k_1)z]}$$

$$\left(\nabla^2 + k_3^2\right)E_{03}e^{-i[\omega_3 t - k_3 z]} = -\frac{\omega_3^2}{c^2}\chi^{(2)}E_{01}\cdot E_{02}\cdot E_{04}^* e^{-i[\omega_3 t - (k_1 + k_2)z]}$$

$$\left(\nabla^2 + k_4^2\right)E_{04}e^{-i[\omega_4 t - k_4 z]} = -\frac{\omega_4^2}{c^2}\chi^{(2)}E_{01}\cdot E_{02}\cdot E_{03}^* e^{-i[\omega_4 t - (k_1 + k_2 + k_3)z]}$$

(10.4)

Details of the further theoretical solution shall not be of interest here, one can read them under [Eng 14] or [Brü 15].

The further consideration resembles the Newtonian mechanics of a spring oscillator with nonlinear deflections x:

$$m\frac{\partial^2 x}{\partial t^2} + k_0 x \begin{cases} = 0 \\ = k^{(2)}x^2 + k^{(3)}x^3 \dots \end{cases} \tag{10.5}$$

- for the upper part in Eq. (10.5), at $=0$, one obtains Hook's law with oscillations of frequency ω and the solution corresponding to (10.2).
- for the lower part in Eq. (10.5) one obtains so-called harmonics; for the term oscillations with frequencies $\omega+\omega=2\omega$ and $\omega-\omega=0\omega$, for oscillations with frequencies $\omega+\omega+\omega=3\omega$ and $\omega+\omega-\omega=\omega$, etc.

In the case of optics, by considering the right-hand part of Eq. (10.3):

- at $=0$ the "linear" or classical optics.
- otherwise, the nonlinear optics; for the term (second order) waves with frequencies $\omega+\omega=2\omega$, for the term (third order) waves with frequencies $\omega+\omega+\omega=3\omega$ and $\omega+\omega-\omega=\omega$, etc.

An illustration of the energetic relations can be obtained in the particle picture: the effect of the nonlinearities can be represented as a "fusion" of two (second order) or three (third order) photons with the same or different frequency to a new photon (Fig. 10.1), where of course the law of conservation of energy of physics must apply.

At high field strength, nonlinear processes can be very effective if, besides the energy conservation, a phase conversation (corresponding to pulse conversation in mechanics) takes place. That means practically:

- for second-order nonlinearities: $(\omega_1 t - k_1 z) \pm (\omega_2 t - k_2 z) = (\omega_3 t - k_3 z)$
  with energy conservation $\hbar\omega_1 \pm \hbar\omega_2 = \hbar\omega_3$ one gets $k_1 \pm k_2 = k_3$

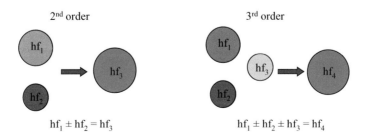

Fig. 10.1 Nonlinearities with particles

- for third-order nonlinearities:$(\omega_1 t - k_1 z) \pm (\omega_2 t - k_2 z) \mp (\omega_3 t - k_3 z) = (\omega_4 t - k_4 z)$
  with energy conservation $\hbar\omega_1 \pm \hbar\omega_2 \mp \hbar\omega_3 = \hbar\omega_4$ one gets.

$$k_1 \pm k_2 \mp k_3 = k_4 \tag{10.6}$$

Since the wavenumber is related to the refractive index ($k = \frac{n\omega}{c_0} = n \cdot \frac{2\pi}{\lambda}$), a wavelength-independent refractive index is a prerequisite for a large effect of the nonlinearity (free of dispersion).

In amorphous optical fibers, due to the lack of symmetry, only nonlinearities of third order and higher play a role. In the following, we restrict ourselves to third-order nonlinearities without further mathematical consideration.

10.1   Nonlinear Effects in Glass Fibers

The number of nonlinear effects in optics is very high; nevertheless, due to the missing symmetry in amorphous glass, only third-order nonlinearities play a certain role [Kra 02]:

- Nonlinear scattering of third order:
  - Stimulated Raman Scattering (SRS),
  - Stimulated Brillouin Scattering (SBS).
- Optical nonlinearities of third order:
  - Four-Wave Mixing (FWM),
  - Self-Phase Modulation (SPM) or Kerr effect,
  - Cross-Phase Modulation (XPM).

10.1.1 Nonlinear Scattering in Glass Fibers

Raman scattering has already been described in Chap. 8. Due to interaction with vibrations in a glass molecule, the energy of incident light will be reduced; thus, we get a higher wavelength. This interaction process is randomized (spontaneously) and we get a *spontaneous* Raman scattering.

If we have high light power (100 mW or more) and if we have light with Raman shift, we can get *stimulated* or *induced* Raman scattering (SRS). It results in some consequences especially for dense wavelength division multiplexing (DWDM): Shorter wavelengths act like a "pump light" and permanently lose power in favor of longer wavelength - the spectrum is changed and we get the so-called *Raman tilt* (Fig. 10.2). This effect can even result even in a crosstalk between different channels.

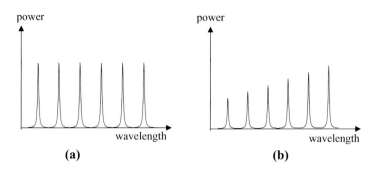

Fig. 10.2 Raman tilt: Power spectrum without (a) and with (b) SRS

Brillouin scattering is very similar to Raman scattering—however, the interaction of light with acoustic vibrations (so-called phonons) takes place. A frequency shift of about 11 GHz toward higher wavelengths occurs. However, since SBS is only effective in the backward direction, an influence on the transmission - if at all - is only to be expected in bidirectional operation.

10.1.2 Third-Order Nonlinearities in Glass Fibers, Four-Wave Mixing, Self-Phase Modulation, Cross-Phase Modulation

The third-order nonlinear effects [Rei 97] are always connected with four participating photons. Frequencies of these photons should be chosen to meet the energy conservation. However the efficiency of this nonlinear interaction is dictated by pulse conservation.

That means for *Four-Wave Mixing* (FWM)

$$hf_{FWM} = hf_1 \pm hf_2 \mp hf_3 \tag{10.7}$$

for all combination of starting frequencies f_1, f_2 and f_3. The "new" frequency f_{FWM} is named ghost. At equidistant frequencies the combination of f_1, $f_2 = f_1 + \Delta f$ and $f_3 = f_1 + 2\Delta f$ is depicted in Fig. 10.3. For example, a "new" frequency f_{123} is a combination of three channels with equal distance Δf:

$$f_{123} = f_1 + f_2 - f_3 = f_1 - \Delta f \tag{10.8}$$

With 3 channels one gets 24 ghost frequencies, 16 of them are identical with initial frequencies f_1, f_2, or f_3.

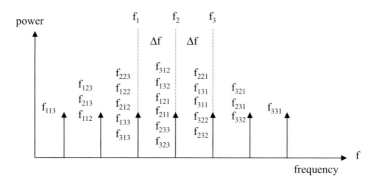

Fig. 10.3 Four-Wave Mixing of 3 frequencies

For a complete consideration we have to include all existing channels used for data transfer (e.g., 40 channels at DWDM). The total number of "ghosts" N_{ghost} can be calculated by

$$N_{ghost} = 0,5 \cdot (N+1)^2 \cdot N \tag{10.9}$$

with N—number of channels. Therefore, with 16 channels we get 2312 ghosts. Ghosts result in a crosstalk—data, e.g., from channel f_1 influence the data stream in channels f_2 or f_3, respectively. To avoid FWM, the total power of *all* channels should be limited—in accordance with the IUT recommendations for the third optical window to maximum of 20 dBm or 100 mW.

At *self-phase modulation* SPM [Rei 97] we get no new frequencies; all changes will occur in the same channel (e.g., at channel with frequency f_1). Thus, energy conservation is

$$h \cdot f_{SPM} = h \cdot f_1 + h \cdot f_1 - h \cdot f_1 = h \cdot f_1 \tag{10.10}$$

On the other hand, at high light power (e.g., in the peak of a pulse) refractive index is changed—for peak of the pulse refractive index is higher (and therefore propagation speed c/n in this part of pulse is lower) compared with the wings. Thus, one gets for SPM a similar situation like in waveguide dispersion (Sect. 2.3.2.3), where the wave in the core propagates at a slower rate compared with the cladding.

At *cross-phase modulation* (XPM) two channels influence each other. At XPM, energy conservation is, for example,

$$h \cdot f_{XPM} = h \cdot f_1 + h \cdot f_2 - h \cdot f_2 = h \cdot f_1 \tag{10.11}$$

Thus, phase of a wave at channel f_2 has an impact on channel f_1. Therefore, data of channel f_2 can be transferred to channel f_1 and vice versa.

10.1.3 Intensity with third-order nonlinear effects in optical fibers.

The most important parameter for the propagation of light in a glass fiber is the refractive index (for *linear* optics, beginning at this point it is denoted as n_0) or group index n_g, respectively. For the transfer of bits in fibers (meaning pulses with a certain power distribution), the refractive index n now depends on peak power $\hat{P}$ (Fig. 10.4):

$$n\left(\hat{P}\right) = n_0 + n_2 \cdot \hat{P} \tag{10.12}$$

At high peak power the refractive index at the needle of the pulse ($\hat{P}_{max}$) is higher than in the edge ($\hat{P}_1$). Thus, the peak part of the pulse propagates more slowly than the wings—the wings "pass" the peak value and we get a nonlinear pulse strain.

It should be noted that such a nonlinear behavior can be observed only if *field strength E of laser light* is comparable with typical *inner-atomic field strength*.

To get an idea of the necessary field strength, one can calculate field strength E from power density S (power per unit of area)

$$S = \varepsilon_0 \cdot \frac{E^2}{2} \tag{10.13}$$

with $\varepsilon_0 = 8.854 \cdot 10^{-12}$ F/m - dielectric constant in vacuum. For 10 mW power and 1 µm beam diameter one can calculate the field strength $E = 6 \cdot 10^7$ V/m.

Field strength E_H in a hydrogen atom between electrons on the First Bohr orbital and the core one can be calculated using Coulomb's attraction by the formula

$$E_H = \frac{\sqrt{3} \cdot e}{4\varepsilon_0 \cdot \pi \cdot r_1^2} \tag{10.14}$$

with $r_1 = 0.5 \cdot 10^{-10}$ m—radius of first Bohr orbital and $e = 1.6 \cdot 10^{-19}$ As—elementary charge. Thus, we get a field strength $E_H = 1{,}1 \cdot 10^{12}$ V/m. Laser field strength is much smaller than inner-atomic field strength, but we have to keep in mind that nonlinearities continue to act in the (very long) fiber.

The nonlinear refractive index is very small ($n_2 \cong 3 \cdot 10^{-20}$ m²/W). In a standard SMF with 9 µm core diameter and laser power of 10 mW, one gets a power density of $1.25 \cdot 10^8$

Fig. 10.4 Changes in peak power and refractive index in time in a pulse

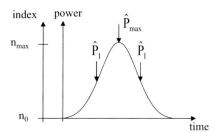

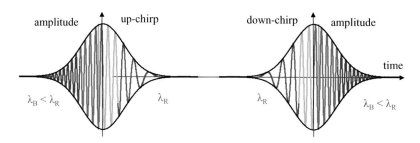

Fig. 10.5   Amplitudes of pulses with up- and down-chirp

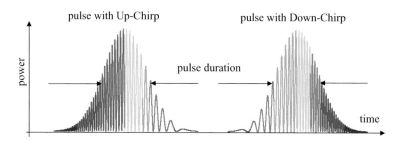

Fig. 10.6   Pulse power with up- or down-chirp

W/m^2. Changes in refractive index are very small (n - n$_0$ ~ 3.8·10^{-12}) and the resulting phase shift is also very small - but the nonlinear effect is increased in a very long fiber.

10.2   Chirp in Glass Fibers

Due to material dispersion, waves of different wavelengths λ_1, λ_2 and λ_3 (or frequencies) arrive at the receiver at different time—the "instantaneous frequency" changes in time (Fig. 10.5). This behavior is like the tweeting or chirping of birds—that's why it is named *chirp*.or as pdf file) The frequency of the "sound" can increase with time (Up-Chirp) or decrease (Down-Chirp).

Thus, the receiver "sees" a bit (or a pulse with a certain pulse shape) as a sequence of sine-like wave with always increasing/decreasing frequency (or decreasing/increasing wavelength). At the receiver one gets the following pictures of up- or down-chirp (Fig. 10.6).

In a receiver one does not measure the amplitude but power P (that means amplitude squared), and one also gets a chirp during the pulse (Fig. 10.6).

The material dispersion which was discussed before behaves like an up-chirp - the "instantaneous frequency" measured at the receiver, increases with time. Note that the spectrum—especially the *line width* (LW) - stays unchanged (Fig. 10.7). Full compensation of the chirp is the key to the generation of extremely short light pulses. All frequencies can be synchronised with a so-called mode locking. The minimum pulse length then corresponds

Fig. 10.7 Line width (LW) of pulses with chirp

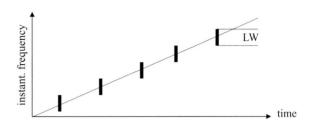

Fig. 10.8 Down-chirp compensation in glass fibers

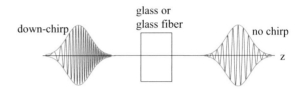

Fig. 10.9 Up-chirp compensation in two gratings

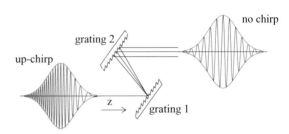

to the bandwidth achievable with a Fourier transformation – these pulses are referred to as bandwidth-limited pulses. For example, a 10 ps pulse at a wavelength of 1 μm corresponds to a bandwidth of 0.2 nm. With this technology, pulse lengths of about 10 femtoseconds (fs) can be achieved in the ultraviolet range of spectrum. Even shorter pulses, down to about 250 attoseconds (as), can be obtained by superposition and oscillation of several fs pulses in noble gases. The 2023 Nobel Prize in Physics was awarded to Pierre Agostini, Ferenc Krausz and Anne L'Huillier for this technology (2023 Nobel Prize in Physics).

There are different possibilities to change the chirp: the chirp can be "compensated", or an initially unchirped pulse can be "re-stamped" to have a certain chirp.

It is possible to change a pulse with chirp to an unchirped pulse or to "impose" a chirp on a pulse without chirp. For example, propagating in glass or glass fiber, an initially down-chirped pulse can be transformed into a non-chirped pulse (Fig. 10.8). Conversely, a chirp-free pulse is transformed into an up-chirped pulse after passing through glass or glass fiber.

Two gratings are able to change an up-chirped pulse to a non-chirped pulse (Fig. 10.9). Or, vice versa, when a non-chirped pulse propagates through gratings, we get a down-chirped pulse. Note that gratings in Fig. 10.9 can be also a fiber with integrated *Fiber-Bragg-Gratings* (see Sect. 8.2.1.2).

In Tab. 10.1 possibilities of chirp generation and compensation by glass and gratings are summarized.

Tab. 10.1 Chirp generation and compensation by glass and grating

Initial situation	Chirp element	Effect of the element
No chirp	Glass	Up-chirp
No chirp	Gratings	Down-chirp
Up-chirp	Glass	More up-chirp
Up-chirp	Gratings	No chirp
Down-chirp	Glass	No chirp
Down-chirp	Gratings	More down-chirp

Fig. 10.10 Theoretical and practical losses in a fiber

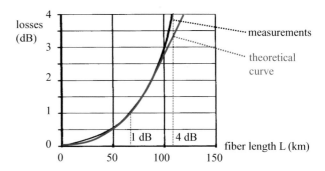

10.3   Polarization Dispersion Management with Nonlinearities

Nonlinearities and their consequences can be either reduced, or completely avoided, or used to compensate dispersion—this is the so-called Polarization Dispersion Management (PDM).

10.3.1 Reduction of Chirp

If we have a light source with chirp, one can calculate the length L_{1dB} (in km) for a 1 dB drop:

$$L_{1dB} = \frac{800}{BR^2} \tag{10.15}$$

BR is the bit rate (in Gbps). For BR = 10 Gbps one gets $L_{1dB} = 8$ km for a chirped pulse. It is possible to increase length L_{1dB} by reducing or switch-off of chirp using special laser methods, e.g., by the so-called *injection locking*. By a suitable external light source (injection) the light source to be used is "forced" to have a *certain* chirp (e.g., up-chirp). This "injected" chirp is then compensated by a glass fiber. For the third optical window, the transmission length L_{1dB} will be

$$L_{1dB} = \frac{6500}{BR^2} \tag{10.16}$$

Taking the same bit rate as before (10 Gbps) we get $L_{1dB} = 65$ km (Fig. 10.10), for 2.5 Gbps even $L_{1dB} = 1000$ km. With increasing losses a (in dB), the length L_{1dB} (in km) increases nonlinearly (theoretical behavior in Fig. 10.10):

$$L \cong \sqrt{a} \qquad (10.17)$$

If the acceptable attenuation is increased from 1 to 4 dB, the transmission length does not increase from 65 to 130 km (corresponding to formula 8.12); we reach "only" 105 km (Fig. 10.10, measurements from [Elr 88]).

10.3.2 Using the Chirp

Now let us consider chirp not as a problem - i.e., in a positive way. Then we can think about how to use chirp in the laser light, and how to compensate this chirp in a fiber. At external modulation, mainly methods of "pre-chirping" are used. An initially unchirped pulse will be re-stamped by external modulation to become a down-chirp. This means that the rising edge contains a shorter wavelength than the falling edge.

Because group index n_g decreases with increasing wavelengths (see term dispersion, e.g., Fig. 3.17), longer wavelengths propagate faster than shorter ones, and we get a "pushing together" of different wavelengths resulting in a pulse shortening (Fig. 10.11). Thus, we can avoid the "smearing" of bits by material dispersion, and due to improved shape, the distinguishability of bits can be improved.

Of course, the most important parameters such as dimensions of chirp, chromatic dispersion, and fiber length should be matched - this method operates only when the length of the fiber is at least roughly the same - i.e., when it is well "matched".

Another method is the so-called Dispersion-Supported Transmission (DST). Simplified, this method operates in the following way (Fig. 10.12):

Frequency modulation is used with 1-bit and 0-bit at different frequencies. Optical power P remains constant, only the frequency of power emission is changed. Let us consider the situation at different points (1–4) of the transmission line.

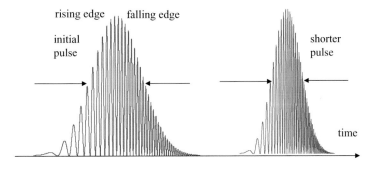

Fig. 10.11 Dispersion compensation by down-chirped pulses

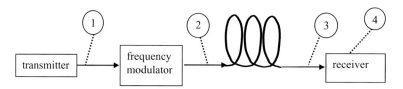

Fig. 10.12 Principle of DST method

The transmitter emits a constant power P (point 1), in the following depicted by a dashed line.

By a frequency modulator, a constant optical power of 1-bit is emitted at a higher frequency (f_H), of 0-Bit at a lower frequency f_L (point 2). Averaged power P remains constant (dashed line in the Figure).

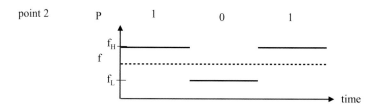

This so-called frequency shift will be arranged in such way that, after running in a fiber of length L, the (faster) 0-Bit arrives at the receiver with a time shift of half pulse duration. Of course, the optical power is changed at receiver (dashed line at point 3) due to the partial overlap of "frequency bits".

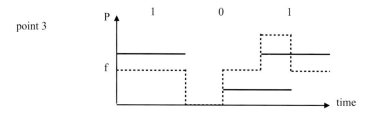

By means of a low-pass filter, for integration in the receiver we retrieve the original transmitter signal (point 4).

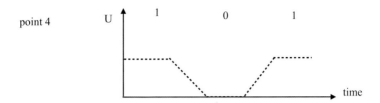

One disadvantage of the DST method is that the frequency pass (shift) should be exactly matched to the fiber length. Experimentally it has been demonstrated that using the DST method, one can bridge about 250 km of fiber length using only amplifiers (without any dispersion compensation).

10.4   Active Compensation of Dispersion

Besides passive dispersion compensation, one can use nonlinear methods to compensate dispersion actively. One active method of compensation is *spectral inversion*. Approximately in the middle of the transmission line, the signal spectrum is "mirrored"(Fig. 10.13). Taking a transmitter signal (bit) of frequency f_1 during propagation in the fiber, one gets chromatic dispersion. In the middle of the transmission line, the signal is coupled by a coupler (C) with a local pump laser of narrow line width of frequency $f_p = f_1 - \Delta f$. The signal (amplified by an EDFA) is mixed in a nonlinear element (NL). Mixing can be performed either in a nonlinear crystal like $LiNbO_3$, or in a semiconductor amplifier SOA, or in a dispersion-shifted fiber (DSF). Thus, we get, e.g., by four-wave mixing an inverted signal at frequency $f_2 = f_p + f_p - f_1 = f_1 - 2\Delta f$ (inverted or mirrored frequency).

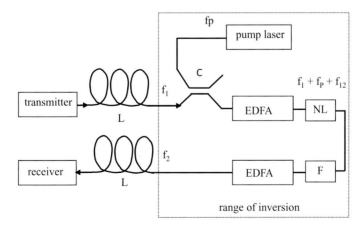

Fig. 10.13 Spectral inversion method

Fig. 10.14 Signal, pump
light and mirrored (inverted)
spectrum [Hul 96]

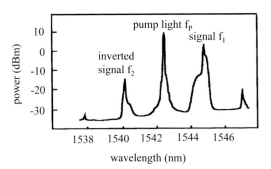

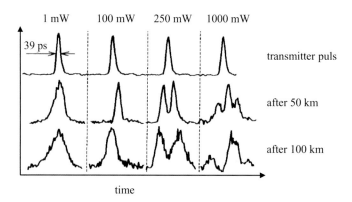

Fig. 10.15 Compensation of dispersion by self-phase modulation [Hul 96]

With a narrow-band filter (F), light at inverted frequency f_2 will be filtered and then
will be amplified in an EDFA. Again, the mirrored spectrum propagates in a fiber, again
we have dispersion but now acting in the opposite direction - dispersion is compensated
(Fig. 10.14).

For example, using this method a dispersion compensated 10 Gbps signal was transmitted over 360 km, and the inversion was performed after 200 km in 21 km dispersion-
shifted fiber. The initial signal power was -0.5 dBm, pump power was 8.9 dBm.

Another possibility of active dispersion compensation is the use of Kerr effect or self-
phase modulation (Fig. 10.15). Under certain circumstances, the pulse smearing (pulse
prolongation) can be compensated by self-phase. The phase shift is very small - but the
effect is increased cumulatively in a very long fiber. Thus, the combination of dispersion
and self-phase modulation can result in dispersion compensation.

As it is depicted in Fig. 10.15, a weak (1 mW) pulse of 39 ps pulse duration experiences a triplication of pulse duration by chromatic dispersion after 100 km - there is
nearly no influence of phase modulation. At high power, we have another situation. The
shape of a 100 mW pulse stays unchanged up to 50 km due to compensation of chromatic dispersion by self-phase modulation - only at longer distances do we find a pulse
broadening due to attenuation and therefore reduction of power. At still higher power

Fig. 10.16 "Dissolving" of
a wave propagating along the
fiber of length L

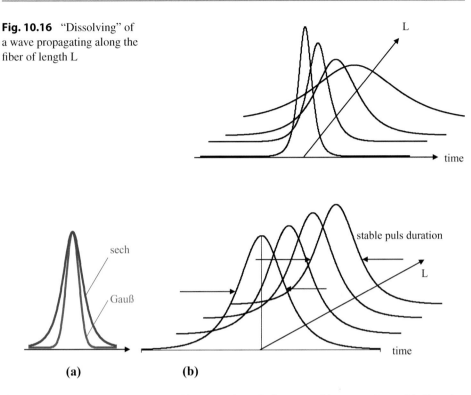

Fig. 10.17 Solitons as inherently stable secant-hyperbolic waves (b), comparison with Gaussian pulse (a)

(250 mW or 1 W), the self-phase modulation dominates, and we get a pulse splitting: the dispersion that results cannot be compensated well.

10.5 Solitons

Nonlinear effects like self-phase modulation could be used brilliantly if power and pulse shape remained stable during propagation in a fiber. But unfortunately, this is normally not the case - we get a pulse broadening by dispersion and amplitude reduction due to attenuation (Fig. 10.16). Peak power and therefore nonlinear effects are reduced permanently.

This situation is generally well-known from water waves - with increasing distance from the source, the amplitude becomes smaller and smaller. But in 1834, the Scotch engineer John Scott Russell made important observations in the English Channel: when a bow wave is created in the front of a ship as it moves through the water, the wave then propagates seemingly lossless along the channel to the water side. Waves propagate with constant speed and without changes in shape along several kilometers. The constant wave shape of these *solitons* (as they were later named) depends on several conditions: from dispersion (in water the water depth) and from nonlinearities (in water the situation at surface area). A special version of soliton waves are tsunami or seismic sea waves - they arise by seaquakes in the ocean and have wavelengths of 10 to 100 km. Obviously these wavelengths are higher than the water depth—that's why at the open sea, the amplitudes of these water waves are typically about 1 m. They propagate constantly and very rapidly and are able to cross the ocean within one day - coming to the coast the water depth is reduced. Because the energy flux is constant, the waves become very high, and we get a flood disaster like in 2004.

In optics and optical communication, we have also done research on solitons - scientists believe to find pulses of stable shape by *combination* of pulse deformation caused by dispersion with the nonlinear Kerr effect. Under certain conditions (power density, pulse duration, wavelength, chromatic dispersion, *and* nonlinear Kerr effect) such a wave with Secant-Hyperbolic-(sech) shape can be very stable - in contrast to Gaussian shaped pulses.

These stable pulses have some other excellent properties: small distortions in shape or in pulse duration will be "self-annealed". The necessary power of soliton pulses is given by an engineering formula deduced from the theory where certain units should be used.

$$P \cong 10 \cdot \frac{\lambda^3 \cdot A_{eff} \cdot D_{chrom}}{\tau^2} \tag{10.18}$$

with P—averaged power (in mW), λ—wavelength (in μm), A_{eff}—effective core area (in μm^2), D_{chrom}—chromatic dispersion (in $\frac{ps}{km \cdot nm}$) and τ—soliton pulse length (in ps).

Exercise 10.1

Which power is necessary to create a soliton in a glass fiber ($D_{chrom} = 20 \ \frac{ps}{km \cdot nm}$, $A_{eff} = 64 \ \mu$m^2) in the third optical window at transfer rates < 100 *Gbps* (bit length $\tau = 10$ ps)?

▶ **Tip**
   Help H10-1 (Sect. 12.1)
   Solution S10-1 (Sect. 12.2)

Exercise 10.2

Which power does one need to create solitons at $\tau = 100$ ps (transfer rate < 10 *Gbps*)?

▶ **Tip**
   Help H10-2 (Sect. 12.1)
   Solution S10-2 (Sect. 12.2)

Optical Networks

11

When the course was set for future communications networks more than 20 years ago, the demand for transmission capacity was still much lower. Voice traffic (with voice signals of 64 kbps), which still dominated at that time, has long since been replaced by digital technologies. Added to this is the data traffic via the Internet and for image transmissions of the highest quality (including broadband TV). The global network was created (Fig. 11.1). It is discussed in detail in [Brü 22]. Today, the major providers earn their money primarily from data traffic.

11.1 Global Networks

Global networks result from the combination of smaller (local) and larger networks. A basic distinction is made between passive networks (PON), in which no signal processing such as amplification takes place, and active networks (AON), which include such signal processing.

The global network (Fig. 11.1) essentially consists of three different parts:

- Local area network (LAN): A local network is usually no more than 500 m long and rarely extends over more than one building complex. It can be designed entirely with electrical lines (eLAN), followed by conversion to optical signals (e ↔ o in the eLAN part, Fig. 11.1). However, radio waves can also be used (WLAN or Wi-Fi). In this case, a router or server converts the radio waves into electrical signals (f ↔ e in the WLAN part, Fig. 11.1). In Germany, this is often connected to a DSL line, which allows transmission rates of up to 100 Mbps via the relatively fast vectoring

Supplementary Information The online version contains supplementary material available at https://doi.org/10.1007/978-3-658-43242-3_11.

V. Brückner, *Elements of Optical Networking*,
https://doi.org/10.1007/978-3-658-43242-3_11

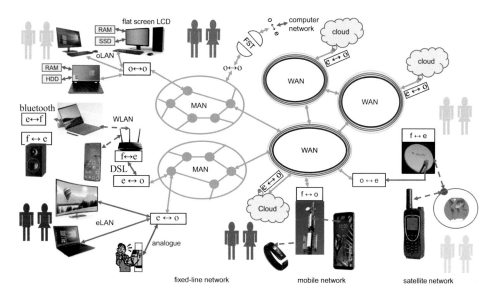

Fig. 11.1 Global network

technology and is one reason why Germany is a "developing country" when it comes to fiber optics in the home (between $f \leftrightarrow e$ and $e \leftrightarrow o$ in the WLAN part, Fig. 11.1 direction MAN). However, LAN can also contain optical fibers (MMF) (oLAN). In this case, it is usually necessary to convert the wavelengths from oLAN (e.g., $\lambda = 750$ nm) to MAN. Furthermore, optical free-space transmission (FST, see Sect. 3.1) is used at least temporary for the coupling of fast computer networks within or between different buildings. In principle, LAN can be laid as a ring or switched network with nodes at least temporary. And don't forget: In many regions of the world, you are lucky if you find an analog telephone at some distance, for example along the railway lines or the transport lines for electrical energy!

- Metropolitan Area Network (MAN): MAN is an interconnection of urban areas that have a high density of offices. It is typically designed as a skeleton (backbone) with SMF and is up to 100 km long.
- Wide Area Network (WAN): WANs are the backbone of global networks. Wide area networks have a fiber optic infrastructure based on SMF. They also include international submarine cables. Distances of several thousand kilometers are not uncommon.

Furthermore, in addition to those shown in Fig. 11.1, there are connections between two points via directional radio using electromagnetic waves - especially if the distance cannot be bridged by electrical or fiber optic cables for some reason. Communication via satellites (satellite radio) can also be seen as a supplement to mobile technology.

Today, there is no longer any fundamental difference between data processing (formerly the domain of the computer) and data communications (formerly belonging to transmission and switching systems). There is hardly any distinction between voice, data, and image communication - all signal types are digital and are treated in the same way [Brü 22].

11.1.1 Transport Networks

Communication networks are classified into transport networks and access networks, as direct link to the subscribers (or user), or as link to local networks (LAN). Connection is realized by optical cross-connectors (OXC). In Fig. 11.2 the communication network is depicted.

The uniform transport network is a fiber optic network, the so-called backbone network (Fig. 11.3). These fiber optic links are already well developed but must be expanded further to ensure sufficient future security. The largest backbone networks in Europe are operated by Deutsche Telekom (220,000 km), Telefonica (40,000 km) and Arcor (33,000 km).

Figure 11.3 shows the network within Germany and the interfaces (connection options) to other networks abroad.

Many WANs together form the global or worldwide network, which is in principle accessible from anywhere on earth and at any time—everyone knows the problems of everyday communication over the Internet!

The large number of optical fibers laid around the globe in the form of submarine cables connecting the continents is also remarkable. Figure 11.4 is intended to provide only a brief overview. An interactive map (https://www.submarinecablemap.com/) shows the current location of submarine cables.

Depending on the length of the fiber optic link, a distinction is thus made between WAN, MAN, and LAN. In the short-haul range (up to about 1 km), fixed links are increasingly being supplemented or displaced by mobile communications technologies. The differences are shown in Fig. 11.5.

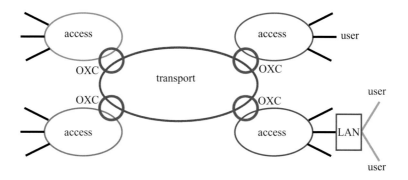

Fig. 11.2 Transport and access networks

Fig. 11.3 Backbone network
of Deutsche Telekom

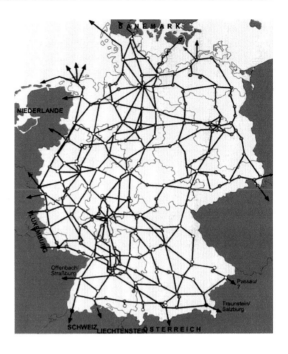

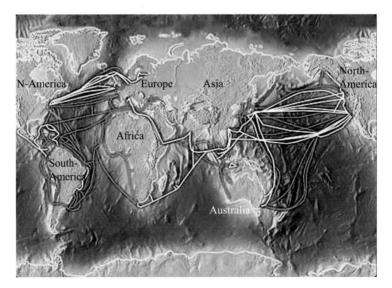

Fig. 11.4 Worldwide networking with glass fibers

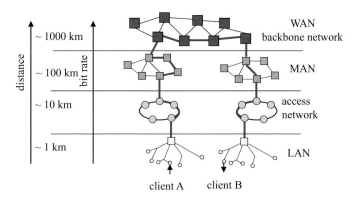

Fig. 11.5   WAN, MAN and LAN

11.1.2 Access Networks, FTTX

The advantage of fiber optic technology is above all a much higher bit rate. Conventional technology with twisted copper wires, e.g., DSL technology, can realize 50 Mbps in the best case. In practice, it is a constant battle between the bitrate promised and the bitrate actually delivered.

In transport networks like WAN or MAN, primarily optical technologies are in use. Access networks represent the connection of the global network to the end customers.

In access networks, increasingly optical technology is used to bring data (Fig. 11.6) to the customers. This is the so-called FTTX technology, where the X usually stands for curb or cabin (C), building (B), or house (H):

- to the curb (fiber to the curb or cabin, FTTC): The fiber ends in a box (in Germany mostly of gray color) up to 500 m away from the customer. The further connection is made via twisted copper wires, e.g., using DSL or VDSL technology. Depending on the distance, up to 50 Mbps downstream and 10 Mbps upstream can be realized.
- to the building (fiber to the building, FTTB): The fiber ends in an apartment building, for example, and is distributed there. Up to 100 Mbps are possible.
- to the home ((fiber to the home, FTTH): The optical fiber ends directly in the end user's house, so only short distances have to be bridged electrically (Ethernet) or wirelessly (WLAN). More than 1 Gbps is therefore possible.

Rarer is FTTD - fiber-to-the-desk: This is the fiber optic connection technology in which the fiber is routed from the local exchange directly to the workplace.

The long distances to the distributor, e.g., in the WAN, are now completely carried out using fiber optic technology. Connection to the clients or customers, so-called "last" mile (in reality it is far less than one English mile, i.e., 1609 m) is still a domain of

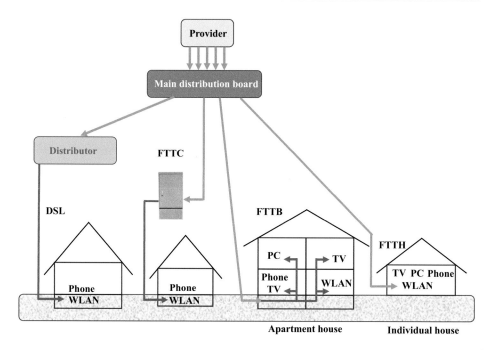

Fig. 11.6 Communication via copper (—) or fiber (—) lines

copper core (twisted pair). Additionally, an increasing part of communication is realized by mobile technologies. For broadband services like image communication and TV, an enormously high data rate is required, which can be realized by glass fiber technology to the client.

FTTH creates the connection between the optical network and the customers' (mostly electrical) networks, such as WLAN or Wi-Fi. The basic idea of an FTTH setup is sketched in Fig. 11.8.

A distinction is made between active and passive optical networks. In an active optical network (AON) there are active elements, such as optical amplifiers. Passive optical networks (PON) are pure power distribution. The passive optical network is like a tree with a trunk and many branches. In practice, FTTH consists of several parts:

Fiber optics along the road to each house, generally as SMF. Often, a composite pipe with up to 24 loose individual pipes is laid in the road or path. This is typically a cable with up to 192 Single-Mode Fibers. To ensure clear identification to the end user, each fiber is color coded. In the author's living area, there is an approximately 2 cm thick fiber optic cable about 60 cm deep in the street. An example of fiber-optical cable is given in Fig. 11.7a.

The cable itself is configured as follows:

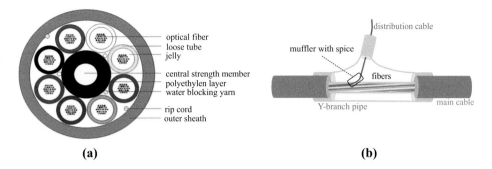

Fig. 11.7 Fiber optic cables in public areas* (a) and y-branches to the house connection (b)
*taken from data sheet Mini Fiber Optic Cable A-DQ2Y nx12/24 of Klaus Faber AG, Saarbrücken

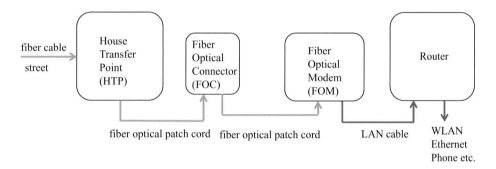

Fig. 11.8 FTTH arrangement (green - fiber, brown - copper cable)

- Up to 8 loose tubes with up to 24 optical fibers each, filled with thixotropic com-
  pound.
- Stranded loose tubes; central strength member made of fiber reinforced plastic.
- Fiber cable strand: Dry, with water-blocking materials.
- Outer sheath: High-density polyethylene with two underlying ripcords.

In order to avoid the costly and time-consuming process of digging up the roads, micro-
excavation is now used. This involves connecting two boreholes at a distance of about
10–20 m with an earth rocket via a horizontal borehole. The installation is carried out
using the blasting technique between the individual ducts. Fiber line lays from public
ground to the cellar (FTTC), to the building (FTTB), or to the house connection room.
The individual SMF are then routed underground by a Y-branch pipe (Fig. 11.7b) in a
distribution cable to the house (FTTH) minimum as a double fiber (two SMF in one
cable). The unused fibers are kept in reserve (so-called dark fibers).
  The entire FTTH deployment consists of:

- House transfer point (HTP) for optical signals: It is a connection of optical lines that can be separated if necessary.
- The fiber optic connection box (FOC for short) in turn: It is the optical connection between the HTP and the optical network termination point (ONT for short) and brings the optical network units (ONUs) closer to the end user's premises (fiber optic subscriber connection). Mostly patch cords with SMF and LC/APC simplex connectors at both ends are used.
- An optical distribution network: It splits and distributes the signal traveling along the passive optical network. This is the so-called fiber optic modem (FOM). The working of a fiber optic modem is quite simple:
  - The Internet signal is transmitted via fiber optic patch cables to the fiber optic modems.
  - The modem transforms the Internet signals into electronic data.
  - This transmission is a full-duplex transmission. That means the data can be transmitted from the Internet source to the electronic device and vice versa.
- Router, which is responsible for forwarding the data via radio waves (WLAN or Wi-Fi) or electrical lines (Ethernet, classic telephone plug) to the end devices.

In some cases (especially in the past few years) FTTH is also concentrated in a single device, which combines all elements of Fig. 11.8.

Definitely in the near future, glass fibers will be installed to the client (FTTH) - in this case one needs to combine different data packets like cable TV (CATV), broadband Internet-TV IP-TV (television, based on the Internet Protocol), analog "classical" telephone POTS (Plain Old Telephone Service), digital, improved wireless telephones DECT (Digital Enhanced Cordless Telecommunications) and PC data services as well as special transmission technologies (especially radio technologies like WLAN) - everything must run as usual. And it must be beneficial and affordable...

11.2   Communication Topologies

One can distinguish between three different communication topologies:

- Peer-to-Peer (Fig. 11.9a):
  Peer-to-Peer connection is a fixed line between 2 clients. It corresponds to the well-known leased line. It can operate in one direction (unidirectional) or in both (bidirectional). The distance between two clients does not play any role - it can be a transatlantic connection with thousands of kilometers as well as a connection between computer and printer in one room. Using multiplex methods, we are able to transmit many channels in one single fiber. That means, Peer-to-Peer connection is not restricted to two clients. Typically, Peer-to-Peer connection contains passive as well as active elements, e.g., optical amplifiers.

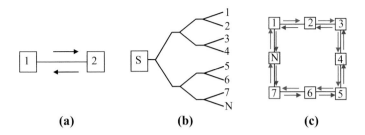

Fig. 11.9 Peer-to-Peer connection (a), distribution system (b), and network (c)

- Distribution system, Peer-to-Multi-Peer connection (Fig. 11.9b):
  In a distribution system, transmitter signals are transmitted to N receivers. Therefore, distribution systems are passive optical networks. Signal power is reduced in each branching point (typically 3 dB per branching point). This is the classic application for a PON. A system of optical branching points is named *tree system*. Data are transmitted only in one direction (downstream). For example, Internet-TV operates with downstream. For interactive communication via cable network or Internet one also needs the other direction (upstream).

Exercise 11.1

In a distribution system of Fig. 11.9b, 3 dB splitters are used. Which power (in mW and dBm) can we expect at each of the eight receivers if the initial power from the transmitter is 10 mW?

▶ **Tip**
   Help H11-1 (Sect. 12.1)
   Solution S11-1 (Sect. 12.2)

- Network, Multi-Peer-to-Multi-Peer connection (Fig. 11.9c):
  In a network, we have a number of stations (clients or terminals) connected with each other. Each station is affiliated to a network node. Thus, we get a *bus system* (Fig. 11.10a). All stations have the *same rights*, i.e., each station can transmit and receive. There are two options for arranging the network nodes with the commonly used communication medium: ring (Fig. 11.10a), or with a mediated network (Fig. 11.10b).

In networks with shared communication medium, there are also several topologies: Bus, ring (Fig. 11.10a) and star. In a mediated network (Fig. 11.10b), each station can connect with another one through nodes. The connection is active only for the duration of the actual communication. This is the playground of the Internet.

Fig. 11.10 Network
with commonly used
communication medium (a)
and mediated network (b)

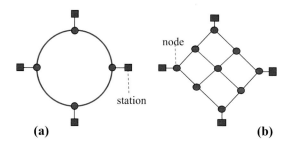

(a)   (b)

11.3   Optical Multiplexing

Multiplex methods are used to increase the transmission capacity in optical fibers, i.e., to
transmit as much data as possible in one channel (e.g., in one fiber). In principle there are
three possibilities:

1. To install more fibers (Space Division Multiplex, SDM).
2. To interleave pulses of different channels in time and to transmit it in shorter time
   slots (Time Division Multiplex, TDM).
3. To transmit signals of different carrier frequencies (wavelengths) in one single fiber,
   this is Optical Frequency Division Multiplex (OFDM) or optical Wavelength Division
   Multiplex (WDM).

Optical frequency multiplex is often denoted as wavelength multiplex, although frequency multiplex seems to be the more correct name; frequency depends only on the
source and is unchanged during propagation through a medium—wavelength is changed
with the speed of light. Furthermore, be careful not to confuse the optical frequency multiplex with the electrical one. Electrical frequency multiplex combines several carrier frequencies on the electrical level.

Figure 11.11 illustrates the possibilities to increase transmission capacity in general.
If we transmit one channel in each fiber and have a cable with, e.g., 100 glass fibers, we
have a space division multiplex (SDM). To better use the capacity of a glass fiber in each
fiber, several wavelengths (e.g., 16 wavelengths) are transmitted—this is the wavelength
division multiplexing (WDM). At each wavelength, one can transmit signal flux in time
slots using time division multiplex (TDM), e.g., 16 time slots with 155 Mbps (STM-1)
are combined by TDM to 2.5 Gbps.

| Exercise 11.2 |

How much is the total transmission capacity of a cable with 100 fibers, if in each fiber
there are 16 wavelengths, and if in each channel (wavelength), STM-1 is realized?

Fig. 11.11 Multiplexing in photonics

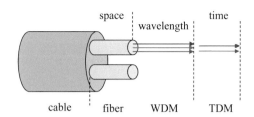

space
wavelength
time

cable   fiber   WDM   TDM

▶ **Tip**
Help H11-2 (Sect. 12.1)
Solution S11-2 (Sect. 12.2)

11.3.1 Space Division Multiplexing (SDM)

In Fig. 11.12 the basic idea of space division multiplexing (SDM) is demonstrated. The electrical signals (each with m bps) of channels number 1, 2, … n, are transmitted into transmitters (TRANS) number 1, 2, … n. The corresponding optical signals are transported by fibers number 1, 2, … n. For n transmitters we need n receivers (REC). Thus, the total transmission capacity is then $n \cdot m$ bps. That is, in SDM technology, each channel has its own fiber.

Exercise 11.3

How many kilometers of glass fiber is needed to transmit 40 channels over 100 km by SDM?

▶ Solution S11-3 (Sect. 12.2)

Fig. 11.12 Idea of space division multiplexing (SDM)

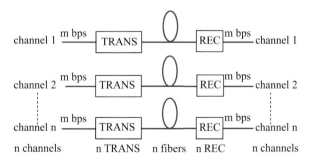

channel 1 — m bps — TRANS — REC — m bps — channel 1

channel 2 — m bps — TRANS — REC — m bps — channel 2

channel n — m bps — TRANS — REC — m bps — channel n

n channels   n TRANS   n fibers   n REC   n channels

11.3.2 Time Division Multiplexing (TDM)

At electrical time multiplex, signals are interleaved in time slots close to each other (ETDM). The so-called Synchronous Optical Networking (SONET) and synchronous digital hierarchy (SDH) were developed in the late 1980s for digital technologies. SONET and SDH are standardized protocols that transfer multiple digital bit streams synchronously over optical fiber using lasers or highly coherent light from light-emitting diodes (LEDs).

There are several recommendations of the ITU-T (e.g., in G.803). The corresponding frame is denoted as synchronous transport module (STM), and the basic frame is STM-1 (Table 11.1). The transmission capacity is increased in each step by a factor of 4, starting from STM-1 (155 Mbps) till (currently) STM-256 (about 40 Gbps).

In Fig. 11.13 n channels with m bps each will be interleaved in time. Multiplexing (MUX) and demultiplexing (DeMUX) run on the electrical level (Fig. 11.13). For transmission, one needs a single optical transmitter, fiber, and receiver. The transmission capacity of a fiber is then $n \cdot m$ bps. If one uses p fibers, we get a total transmission capacity of $p \cdot n \cdot m$ bps.

Exercise 11.4

Forty channels in a fiber are multiplexed by TDM with STM-64. What is the transmission capacity?

▶   Solution S11-4 (Sect. 12.2)

Tab. 11.1 Synchronic transport modules and corresponding bit rates

STM step	Bit rate
STM-256	39,813.12 Mbps
STM- 64	9953.28 Mbps
STM- 16	2488.32 Mbps
STM- 4	622.08 Mbps
STM- 1	155.52 Mbps

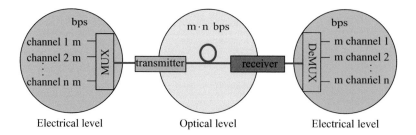

Fig. 11.13 Time multiplex

Just as scaling up from 2.5 Gbps (STM-16) to 10 Gbps (STM-64) resulted in many technical problems, the recent change-over to 40 Gbps (STM-256) which is now commercially available is much more complicated and challenging. Whereas 100 Gbps have been demonstrated already, 160 Gbps (STM-1024) are still being tested in the laboratories. At these extremely high bit rates, dispersion limits the maximum fiber length. Like in 40 Gbps systems, researchers and engineers have to check carefully whether all requirements for the fiber can be achieved.

Another possibility to transmit high data rates is optical time division multiplexing (OTDM). At OTDM different optical pulse sequences (bit patterns) are shifted in time in such a way that the pulses of one signal are placed in the free space between the other signals. An example of a 160 Gbps data transfer is given in Fig. 11.14. A laser emits very short (e.g., 25 ps) pulses, and the time between pulses is, e.g., 100 ps. In this example we want to transfer four channels with 10 Gbps each. By four independent modulators, laser light is modulated by data to be transmitted. Besides modulators, we need to introduce a fixed time shift of the pulse with respect to the other pulses (optical delay). That means that the basic idea of OTDM is similar to electrical time multiplexing - nevertheless it operates completely on the optical level.

Resulting bit sequences should be shifted in time at a quarter period and coupled into a fiber. The time shift (delay) can be achieved by delay lines like different fiber lengths. To achieve a 25 ps delay, the fiber length should have a 5 mm difference with respect to the others. As sum bit rate we get 40 Gbps. At the receiver side pulses will be demultiplexed in a corresponding way to 4 channels with 10 Gbps. One can also use delay lines. Finally, the optical pulses are transformed to electrical pulses by four receivers.

Exercise 11.5

Ten channels should be interleaved by OTDM with a 100 ps frame. What is the maximum pulse duration? Which bit rate per channel can be achieved?

▶   **Tip**
    Help H11-5 (Sect. 12.1)
    Solution S11-5 (Sect. 12.2)

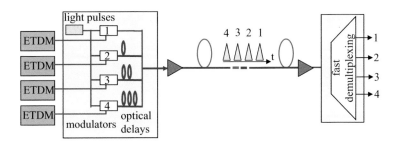

Fig. 11.14   Optical time division multiplex (OTDM)

For the next generation of Internet, bit rates of 160 Gbps or more are under study. By 160 Gbps one can transmit, e.g., 2.5 million phone calls - all at the same time and in one single channel.

In the following we would like to discuss the question of how to increase the transmission capacity in a fiber without changing the bit rate of a channel.

11.3.3 Wavelength Division Multiplexing (WDM)

Optical frequency division multiplexing is usually referred to as wavelength division multiplexing, although frequency division multiplexing is the more appropriate term, since the wavelength changes with the speed of light, the frequency depends solely on the source and generally remains constant when passing through matter. Furthermore, optical frequency division multiplexing must not be confused with electrical frequency division multiplexing, in which multiple carrier frequencies are combined in the electrical plane.

In wavelength division multiplexing (WDM), transmission channels with different wavelengths are combined into one optical fiber with as little loss as possible (MUX); in demultiplexing (DEMUX), they are separated again. In general, the optical principles are the same because of the reversibility of the optical path - but demultiplexing is usually more complicated. The methods described below can therefore be used for both multiplexers and demultiplexers. Multiplexers and demultiplexers are based on filters that can decouple individual wavelengths. Furthermore, optical isolators and circulators are used, these elements are combined in optical add-drop multiplexers and optical cross-connectors.

In the multiplexer (MUX), light from n different channels (here: wavelengths) is coupled together into one optical fiber; in the demultiplexer (DEMUX), they are separated again (Fig. 11.15).

Exercise 11.6

At 16 wavelengths 40 Gbps per channel are transferred. What is the total transmission capacity of the fiber?

▶   **Tip**
    Help H11-6 (Sect. 12.1)
    Solution S11-6 (Sect. 12.2)

Fig. 11.15 Multiplexer and demultiplexer

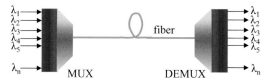

Coupling multiple wavelengths into a fiber increases the transmission capacity. The electrical signals (e.g., consisting of TDM signal streams) each modulate an optical transmitter, the optical signals are combined in a multiplexer and coupled into a fiber (Fig. 11.16). At the end of the transmission path, the wavelength channels are separated in a demultiplexer (optical filter), and each is converted to the electrical level in a receiver.

Per fiber, the capacity is now $n \cdot m$ bits per second. Another advantage of WDM technology is the flexibility with which channels can be switched on or off as needed. All channels in the system do not have to be established immediately, but only when the need arises. Optical multiplexing technology can be used to transmit different data formats such as Gigabit Ethernet, IP, SDH, voice and video, or other services in desired combinations and at different bit rates. WDM technology is protocol neutral.

Exercise 11.7

At 16 wavelengths 40 Gbps per channel are transferred. What is the total transmission capacity of the fiber?

▶ **Tip**
Help H11-7 (Sect. 12.1)
Solution S11-7 (Sect. 12.2)

Various WDM technologies were used in the past and are used today.

11.3.3.1 WWDM
Wide Wavelength Division Multiplexing (WWDM) operates with a small number of channels with large channel distance (>50 nm), e.g., one channel in the O band and one channel in the C-band. This version is cost-saving and useful, especially for 10 Gbps Ethernet.

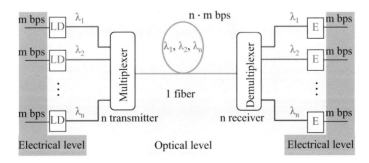

Fig. 11.16 Wavelength division multiplexing (WDM)

11.3.3.2 CWDM

Coarse Wavelength Division Multiplexing (CWDM) is a WDM technology with large (20 nm) channel distances operating in the optical spectrum. Thus, as transmitters one can use LEDs or uncooled lasers, instead of cost-intensive temperature-controlled lasers. CWDM systems are used at present in the "classical" optical windows in the 1310 and 1550 nm range. CWDM permits the transfer of up to 18 independent channels in one fiber pair resulting in a significant cost reduction. Therefore, distances of up to 120 km and data rates of up to 10 Gbps are possible using CWDM.

11.3.3.3 *DWDM*

Dense Wavelength Division Multiplexing (DWDM) or Ultra-DWDM (UDWDM) is an optical wavelength multiplex with very high channel density in the 1550 nm range; the channel distance is between 0.8 nm and 1.6 nm.

The ITU-T recommendation G.692, the so-called ITU-Grid (Tab. 11.2), regulates the wavelength and channel distances. More than 100 channels can be transmitted in a glass fiber pair over distances up to 720 km.

The wavelength (channel) distance depends on the application:

In ITU-T G.671, G.694.1 and G.694.2 the wavelength of the channels is fixed. For DWDM the channel distances are fixed as 12.5 GHz, 25 GHz, 50 GHz, and 100 GHz starting with the frequency 193.1 THz (corresponding to a wavelength of 1553.60 nm).

Therefore, we obtain

$$193.1\,\text{THz} + n \cdot 0.0125 \quad \text{(UDWDM)}$$
$$193.1\,\text{THz} + n \cdot 0.025 \quad \text{(UDWDM)}$$
$$193.1\,\text{THz} + n \cdot 0.05 \quad \text{(DWDM)}$$
$$193.1\,\text{THz} + n \cdot 0.1 \quad \text{(DWDM)}$$

with integer values for $n = \cdots - 3, -2, -1, 0, +1, +2, +3 \ldots$

In Fig. 11.17 one can see four channels with a distance of 100 GHz (corresponds to $\Delta\lambda \cong 0.75\,\text{nm}$). In accordance with ITU-T recommendations, it is not allowed to use the whole channel width; there should be a safety clearance of about 20% to avoid any

Tab. 11.2 Overview on WDM technologies

Standards after ITU-T	WWDM	CWDM	DWDM/UDWDM
Channel distance	≥ 50 nm	20 nm, >2.5 THz	<1000 GHz
Wavelength bands	O, C	O, E, S, C, L	C, L
Number of channels	2	18	100
Applications	Passive Optical Networks	Metro, short distances	Large distances
Costs	low	low	high

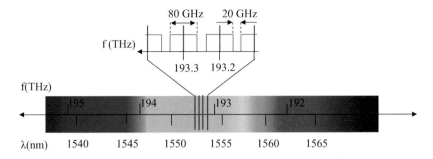

Fig. 11.17 Frequencies or wavelengths and channel structure (4 channels) for DWDM ($\Delta f = 100$ GHz)

crosstalk between channels. As one can see in Fig. 11.17, from 100 GHz we can use only 80 GHz ($\cong 0,6$ nm) as channel width. To use the permitted channel width, one needs to modulate the central frequency (e.g., 193.2 THz) by 40 GHz. Roughly it corresponds to a transmission rate of 40 Gbps; in total we get for four channels a transmission rate of 160 Gbps. There are very strong requirements concerning the stability of laser wavelength: At most $\pm 2.5\%$ of channel distance (in the case of Fig. 11.17 it is 2.5 GHz or 0.02 nm) are permitted as deviation.

Exercise 11.8

How much is the maximum transmission rate, if 80 channels should be transmitted by UDWDM (channel distance 25 GHz) under strict observation of 20% safety clearance? Let us assume that the bit rate and frequency are equal.

▶ **Tip**
Help H11-8 (Sect. 12.1)
Solution S11-8 (Sect. 12.2)

Already in 2004 Alcatel, Deutsche Telekom, and France Telecom reported the transmission of 170 Gbps per channel in 430 km standard SMF. With eight channels, about 1.28 Tbps were transmitted in the region of Marseille, France.

For practical use it is important to have attenuation (fiber) and amplification (e.g., in EDFA) independent of wavelength, if possible. In Fig. 11.18, one can see the attenuation and dispersion of some commercially available fibers in the DWDM and CWDM ranges. As one can see, the attenuation in the c-band is nearly identical, e.g., for 40 channels with a distance of 100 GHz (i.e., a total of about 30 nm).

In CWDM systems the typical channel distance is 20 nm; at WWDM the distance is more than 50 nm, e.g., two channels at 1310 nm and 1550 nm. While in DWDM systems, emission of a very narrow spectral line of transmitter and temperature stabilization are required, for CWDM, uncooled DFB or Fabry-Perot lasers can be used. The shift of

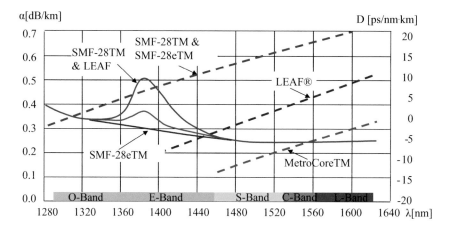

Fig. 11.18 Wavelength bands, attenuation, and dispersion of several fibers of the company Corning Glass (other producers offer similar fibers)

spectrum in a Fabry-Perot laser stays within the limits and we can keep the channel distance of 20 nm.

Figure 11.19 shows the position of channels at CWDM. In accordance with ITU G 694.2 recommendations, the wavelength between 1271 and 1611 nm can be used, i.e., using fibers without water peak ("low water peak" or zero water peak ZWP fiber) one can use 18 channels.

In Fig. 11.20 channels of a DWDM system in the C-Band with 200 GHz channel distance together with one CWDM channel are depicted. In CWDM systems, temperature between 0° and 70° C is allowed; furthermore, one can use standard laser diodes with deviations from the nominal wavelength of ±3 nm. It results in a significant reduction for investment and operation costs.

Another advantage of WDM technology is that we need only *one* optical amplifier for long-distance transmission. Using "classical" (non-optical) amplifiers one needs after

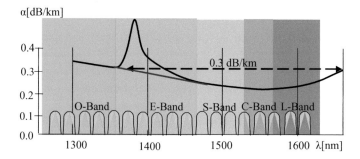

Fig. 11.19 Position of CWDM channels

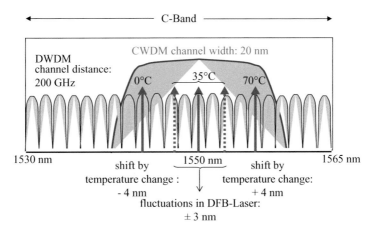

Fig. 11.20 Comparison between DWDM and CWDM

demultiplexing the wavelength channels opto-electrical transformation is needed, afterwards signals will be refreshed, electro-optically transformed, and coupled into the outgoing fiber. Additionally, one has to compensate dispersion after a certain fiber length by a dispersion compensating fiber. Note that full dispersion compensation seems to be not possible due to different propagation speeds of different wavelengths.

In Fig. 11.21 one can see the schema of a multichannel WDM system with optical amplifiers. All wavelengths in the range between 1525 and 1565 nm will be amplified in *one single* amplifier. Only at the fiber length where maximum dispersion is achieved, will the dispersion be compensated by regeneration on the electrical level. In the example, one can see a 40-channel system; by the optical amplifier (instead of a "classical" non-optical amplification) we do not need one multiplexer, one demultiplexer, 40 receivers, and 40 laser diodes including the controlling and temperature equipment as well as

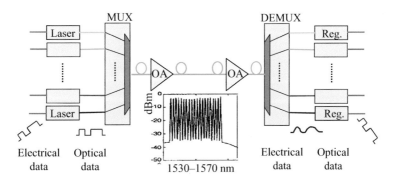

Fig. 11.21 Multichannel WDM system with optical amplifiers (OA: optical amplifier, MUX: multiplexer, DEMUX: demultiplexer, Reg.: regenerator)

an electrical repeater. It is no problem to add a channel: we only have to add a laser transmitter and a receiver - the multiplexer and demultiplexer must be upgradable.

In Fig. 11.22 the advantage of WDM with optical amplifier compared with SDM is illustrated. A system with 8 channels and 3 sections of road with 120 km each, operating in the C-band is assumed. The maximum transmission length is 120 km - after that distance, signals must be either regenerated or amplified. Of course, WDM and TDM can be combined in one transmission system.

In the case of a pure SDM system with a regenerator (upper part in Fig. 11.22), one needs 3×8 fibers per channel, i.e., a total of 2880 km, 8 optical transmitters, 8 receivers, and 16 regenerators. In the case of a WDM system (lower part in Fig. 11.22), we need 8 transmitters and 8 receivers as well, one MUX and one DEMUX, 3×120 km fiber $= 360$ km fiber, and two optical amplifiers. If we want to increase, e.g., the data rate in an SDM system, one has to substitute transmitters and receivers as well as all regenerators (the most expensive components of the system); in a WDM system, one has to substitute transmitters and receivers only. One can also consider whether an additional wavelength can meet the demands. Without optical amplifiers in a WDM system we have to insert a repeater for each wavelength. Herein one can see the high flexibility of WDM technology in connection with optical amplification.

11.3.3.4 CDMA

Code division multiplexing (CDM or Code Division Multiple Access, CDMA) is a multiplexing method that enables the simultaneous time transmission of different user data streams on a common frequency range. In contrast to the classical multiplex procedures such as frequency division multiplex and time division multiplex, with code division

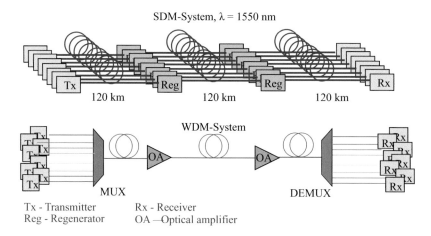

Fig. 11.22 Comparison of SDM with regenerators and WDM with optical amplifiers, total length 360 km, upper part: $\lambda = 1550$ nm, partial length 120 km each; lower part: λ in C-Band, partial length 120 km each

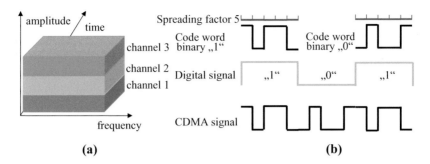

Fig. 11.23 Principle of CDMA (a) and example of spreading the channel in use (b)

multiplex the individual data streams are superimposed both in the frequency range and in the time range (Fig. 11.23a). With CDMA, the focus is on the separation and differentiation of different and parallel transmitted traffic streams over a shared frequency band. The main characteristic of the shared frequency band is that it has a larger bandwidth than the user data stream occupies.

CDMA has several advantages over other technologies: more flexibility in data rates, higher transmission capacity, and higher quality of service in terms of network requirements and support for data bundling.

The coding is based on a spreading of the user data channel. At the input, the individual bits of a narrowband useful signal are replaced by longer bit combinations, the code words. If one bit is replaced by a bit combination of, for example, five bits (Fig. 11.23b), then a spreading by a factor of 5 is achieved. A codeword for the binary "1" and the binary "0" belongs to each spreading factor (Fig. 11.23b). These code words are added to the digital signal, resulting in the CDMA signal (Fig. 11.23b), which is transmitted in the following. At the output of the transmission, the code words are subtracted, and the original digital signal is obtained again. Although CDMA requires a higher transmission bandwidth, the transmission channel can be used for more user channels at the same time. The data of the individual users can be clearly distinguished from each other in the transmission channel.

The main applications of CDMA are currently in the field of digital signal transmission in third-generation mobile radio networks (CDMA2000). Other areas of application for CDMA are the satellite navigation systems like the American Global Positioning System (GPS) or the European Galileo.

11.4   WDM Systems

In this chapter we would like to discuss some ideas about designing WDM systems. Attenuation, dispersion, nonlinearities, and noise have an essential influence on the transmission.

In accordance with ITU-T G.681, we want to consider several scenarios. In Fig. 11.24 three differently standardized cable lengths are depicted. The averaged attenuation per length unit is equivalent to 0.275 dB per km distance. Here attenuation by splices and the so-called system reserve are already included; typical fiber attenuation at 1550 nm is 0.23 dB/km. At Long Haul distance, a maximum of seven amplifiers with 22 dB amplification are permitted - therefore, 640 km transmission length can be achieved. At Very Long Haul distance, a maximum of five partial lengths and four optical amplifiers with 33 dB amplification are permitted - therefore, 600 km transmission length can be achieved. Extension of transmission length using more amplifiers is forbidden due to the noise in amplifiers which will also be amplified. Noise in optical amplifiers results from spontaneous emission. It will be amplified together with the signal along the whole transmission line. The higher the amplification, the higher is the amplified spontaneous emission (ASE). In Ultra Long Haul distances, amplifiers cannot be used because there is no amplifier with 44 dB amplification! After the distance given in Fig. 11.24, signals must be regenerated. In conclusion one can say that the longer the partial lengths, the lower the number of line sections.

There are some other points to keep in mind:

a) The existence of dispersion; dispersion can be compensated by DCF completely only for one single wavelength.
b) Polarization mode dispersion - especially at transmission rates above 10 Gbps.
c) Power budget.
d) Position of channels.

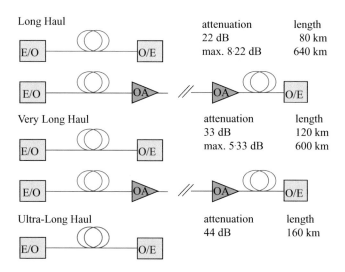

Fig. 11.24 Optical transmission lines of different lengths and attenuation

In accordance with ITU recommendations, the maximum length (of the fiber lengths) is determined by the channel density, the bit rate, and the selection of channels. The density of channels is determined by the discriminatory power of optical filters as well as by the spectral width of transmitters. Each channel has a certain optical power. If the total power is too high, we will have nonlinear effects which can affect the transmission negatively.

We can draw the following conclusions:

- The more channels, the lower the maximum distance.
- The higher the bit rate, the lower the maximum distance.

The design of DWDM systems is a very complex task; typically, it is performed by numerical simulations.

Help and Solutions

12

12.1 Help

Chapter 1:
H1-1: First call the help.

Chapter 2:
H2-1: Use Eqs. (2–1) and (2–3).
H2-2: Substitute Eq. (2.9) into (2.8).
H2-3: Substitute Euler's identity into Eq. (2.10).
H2-5: Use Eq. (2–12).
H2-6: Use Eq. (2–13).
H2-7: Use Eq. (2–15)
H2-8: Use equation $E = P \cdot \tau_P$.
H2-9: Use solution of task 2–8, substitute into (2.17).

Chapter 3:
H3-1: Use Fig. 3–2.
H3-2: Use Eqs. (3.13), (3.14) and (3.15).
H3-3: Use Eqs. (3–15) and (3–14).
H3-4: Use Fig. 3–7.
H3-5: NA from Eq. (3.23), n_{core} from Eq. (3.13) and with $n_{cladding} = 1.445$.
H3-6: Use Eq. (3.24).
H3-7: Use Eq. (3.28).
H3-8: See Tab. 3–2 and Eq. (3.29).
H3-9: See Tab. 3–2.
H3-10: See numbers in the text.

© Springer Fachmedien Wiesbaden GmbH, part of Springer Nature 2024
V. Brückner, *Elements of Optical Networking*,
https://doi.org/10.1007/978-3-658-43242-3_12

H3-11:   See text.
H3-12:   Use Eq. (3.35).
H3-13:   Use Eq. (3.36).
H3-14:   Use Eqs. (3.36) and (3.38).
H3-15:   Use Eq. (3.46).
H3-16:   Use Eq. (3.47).

Chapter 4:
H4-1:   Use formulae for a_d.
H4-2:   Use formulae for a_d.
H4-3:   Use formulas for a_d and a_g.
H4-4:   Use formulae for a_o.
H4-5:   Use formulae for a_α.
H4-6:   Use formulae for a_L.
H4-7:   Losses are adding up.
H4-8:   Use formulae for a_w.
H4-9:   Use formulae for a_w.
H4-10:   Use formulae for a_w.
H4-11:   Use formulae for a_w.
H4-12:   See text chapter 3.2.2.2.

Chapter 5:
H5-1:   Use Eq. (5.4).
H5-2:   See chap. 5.2.1.
H5-3:   See chap. 5.2.2.
H5-4:   See formulas in: system $Ga_{1-x} Al_x$ As on GaAs.
H5-5:   See formulas in: System InGaAsP on InP.
H5-6:   See formulas in: System InGaAsP on InP.
H5-7:   See formulas in: System InGaAlAs on InP (cubic equation).
H5-8:   See formulas in: System InGaAsP or InGaAlAs on InP.
H5-9:   From system InGaAsP on InP (see formulas).
H5-10:   See text chapter 5.2.2.
H5-11:   See formulas to quadratic recombination.
H5-12:   See formulas to Auger recombination.
H5-13:   Consider mobility of electrons and holes!
H5-14:   See Fig. 5–9.
H5-15: $\Delta f(L_M) = c/2L$.
H5-16:   How often does $\Delta f(L_M)$ "fit" into Δf(fluorescence)?
H5-17:   Use Fresnel's formulas!
H5-18:   See Figs. 5–13 and 5–16.
H5-19:   See text!
H5-20:   See text!

H5-21:   See text!
H5-22:   See text!
H5-23:   See text!

Chapter 6:
H6-1:   Draw 3 dB drop in the figure!
H6-2:   Use Eq. (6.3).
H6-3:   See Fig. 6.17.

Chapter 7:
H7-1:   See Eq. (2.10).
H7-2:   See chapter 7.4.
H7-4:   See chapter 7.4.3.
H7-5:   See chapter 7.4.4.
H7-7:   Calculate photon energy hf.

Chapter 8:
H8-1:   Use Eq. (8.2).
H8-2:   Use Eq. (8.3).
H8-3:   Use Fig. 8–33!
H8-4:   Use Fig. 8–39!
H8-5:   See text below Fig. 8–39!
H8-6:   See text in chapter 8.4.1.4.
H8-7:   Note pump capacity limit!
H8-8:   Spontaneous Raman scattering not in phase, stimulated scattering in phase.
H8-9:   See text in chapters 8.4.1 and 8.4.2.
H8-12:   Take care of dispersion types.
H8-13:   See explanation in the text.
H8-14:   See explanation in the text.
H8-15:   Look for possibilities of averaging!

Chapter 9:
H9-1:   See text.
H9-2:   See text.
H9-3:   See text
H9-4:   See Eq. (9.4).
H9-5:   See text.

Chapter 10:
H10-1:   Use Eq. (10.14).
H10-2:   Use Eq. (10.14).

Chapter 11:

H11-1: 10 dBm = 10 mW.

H11-2: See text above the exercise.

H11-4: See text above the exercise.

H11-5: STM1 = 155 Mbps.

H11-6: See text above the exercise.

H11-7: See text above the exercise.

H11-8: See text above the exercise.

12.2   Solutions

Chapter 1:

S1-1: This is the solution

Chapter 2:

S2-1: $2 \cdot 10^{14}$ Hz and 0.8 eV

S2-4: Flat, unlimited waves

S2-5: 20 dBm

S2-6: 3 dB

S2-7: 4.6 dB and $5 \cdot 10^{-5}$ dB

S2-8: 10^{-10} Ws

S2-9: $6.25 \cdot 10^{8}$ photons

Chapter 3:

S3-1: $SiO_2 + 13\%$ GeO_2

S3-2: $\varphi_A = 9.9°$, $NA = 0, 17$, $\Delta = 0.11$

S3-3: $n_{core} = 1.460$ at $n_{cladding} = 1.445$

S3-4: It becomes less

S3-5: $n_k \leq 1-464$

S3-6: no, $\lambda_c = 1176$ nm

S3-7: About 12 dB

S3-8: (a) 0.17 dB/km (b) 0.86 dB/km

S3-9: 2.475 dB

S3-10: 4 dB/km

S3-11: Group index and group velocity

S3-12: 50 ns/km and 20 m

S3-13: 250 ps/km and 4 km

S3-14: 260 ps/km (MMF) and 17 ps/km

S3-15: 1110 km (a), 70 km (b) and 4.3 km (c)

S3-16: 25 km and 400 km

Chapter 4:

S4-1:	1.94 dB
S4-2:	14.9 dB
S4-3:	2.17 dB
S4-4:	0.15 dB
S4-5:	0.70 dB
S4-6:	0.05 dB
S4-7:	0.90 dB
S4-8:	0.639 dB
S4-9:	0.013 dB
S4-10:	0.214 dB
S4-11:	0.011 dB
S4-12:	1: 1 splitting to gates

Chapter 5:

S5-1:	(a) 12 per cm, (b) 0.12 per cm
S5-2:	Radiative transition unlikely
S5-3:	Adaptation of lattice constant necessary
S5-4:	$x=0.027$
S5-5:	$x=0.28, y=0.62$
S5-6:	$x=0.41, y=0.89$
S5-7:	$x=0.168, y=0.302$
S5-8:	See task 5–7
S5-9:	For 1522 nm: $x=0.4045$ and $y=0.8694$; for 1523 nm: $x=0.4050$ and $y=0.8705$
S5-10:	Lattice adjustment necessary
S5-11:	300–50 ns or 30–5 ns.
S5-12:	800–160 ps or 8–1.6 ps
S5-13:	Hole transport much slower than for electrons
S5-14:	Too less wavelength
S5-15:	300 GHz
S5-16:	20
S5-17:	$0.2 \cdot 10^{-5}$
S5-18:	Combine Figs. 5–13 and 5–16, reason: resonator
S5-19:	Existence of threshold
S5-20:	Power reduction with temperature increase or aging
S5-21:	Circular (better for fibers) or elliptical beam cross section
S5-22:	Current control, three-part power supply
S5-23:	Circular beam cross section, better focusable in MMF/SMF

Chapter 6:

S6-1:	About 1.3 or 1.5 GHz
S6-2:	Threshold, relaxation; no – LED has a linear characteristic

S6-3:   At gate 3: "0" without voltage, "1" with voltage; at gate 4 inverted

Chapter 7:
S7-1:   3.2 µW
S7-2:   Homogeneous distribution of the noise over frequencies
S7-3:   See text, chapter 7.4.1
S7-4:   Heat movement, cooling of receiver
S7-5:   All types of noise + multiplication noise
S7-6:   See Eqs. (7.13) and (7.18)
S7-7: $2.7 \cdot 10^{-18}$ Ws

Chapter 8:
S8-1:   2 kV
S8-2:   1 kV/cm
S8-3:   Only in third optical window
S8-4:   The lowest gain is decisive
S8-5:   Complementary power distribution at the input or ZBLAN
S8-6:   Forward: for small signal amplification; backward: for strong signals
S8-7:   Total permitted power in the fiber is limited
S8-8:   Stimulated Raman scattering
S8-9:   Advantage: flexible λ; disadvantage: high power necessary
S8-10:   SOA has no mirrors
S8-11:   To avoid laser action
S8-12:   Negative waveguide dispersion
S8-13:   5.5 km
S8-14:   Transit time differences for different wavelengths
S8-15:   Averaging over the correlation function

Chapter 9:
S9-1:   0.5 dB/km
S9-2:   Plugs have a needle-shaped tip
S9-3:   Multiple splices or a very bad splice
S9-4:   1.12 GHz
S9-5:   Very poor quality

Chapter 10:
S10-1:   432 mW
S10-2:   4.32 mW

Chapter 11:
S11-1:   1.25 mW
S11-2:   4000 km

References

[Brü23]   [Brü 23]Brückner, V.: Elemente optischer Netze: Grundlagen und Praxis optischer Datenübertragung. 3. Auflage. Springer-Vieweg Verlag 2023.

[Brü22]   [Brü 22]Brückner, V.: Globale Kommunikationsnetze: Über Digitalisierung, elektromagnetische Wellen, Glasfasern und Internet. Springer-Vieweg Verlag 2022.

[Zie08]   [Zie 08]Ziemann, O.; Krauser, J.; Zamzow, P. E.; Daum, W.: POF-Handbook (Optical Short Range Transmission Systems). Berlin, Heidelberg, New York: Springer 2008.

[Sal08]   [Sal 08]Saleh, B. E. A.; Teich, M. C.: Grundlagen der Photonik. Weinheim: Wiley Verlag GmbH & Co. KGaA 2008.

[Bör89]   [Bör 89]Börner, M.; Trommer, G.: Lichtwellenleiter. Stuttgart: Teubner-Verlag 1989.

[Ped08]   [Ped 08]Pedrotti, F.; Petrotti, L.; Bausch, W.; Schmidt, H.: Optik für Ingenieure. Berlin, Heidelberg, New York: Springer-Verlag 2008.

[LTU87]   [LTU 87]Grundlagen der Optoelektronik. Bremen: L.T.U.-Vertriebsgesellschaft 1987.

[Vog02]   [Vog 02]Voges, E.; Petermann, K. (Hrsg.): Optische Kommunikationstechnik. Berlin, Heidelberg, New York: Springer Verlag 2002.

[Rei05]   [Rei 05]Reider, G. A.: Photonik – Eine Einführung in die Grundlagen. Springer-Verlag, Wien 2005.

[Brü98]   [Brü 98]Brückner, V.: Messtechnik für faseroptische Systeme. Unterrichtsblätter der Deutschen Telekom Nr. 7, S. 294–307; 1998.

[Blu95]   [Blu 95]Bludau, W.: Halbleiter-Optoelektronik. München, Wien: Carl Hanser Verlag 1995.

[Har98]   [Har 98]Harth, W.; Grothe, H.: Sende- und Empfangstechnik für die optische Nachrichtentechnik. Stuttgart, Leipzig: Teubner-Verlag 1998.

[Wag98]   [Wag 98]Wagemann, H.-G.; Schmidt, A.: Grundlagen der optoelektronischen Bauelemente. Stuttgart: Teubner-Verlag 1998.

[Ker83]   Kersten, R. Th.: Einführung in die Optische Nachrichtentechnik. Berlin, Heidelberg, New York: Springer Verlag 1983

[Hil95]   [Hil 95]Hillmer, H.; Grabmaier, A.; Zhu, H.-L.; Hansmann, S.; Burkhard, H.: Continously chirped DFB gratings by specially bent waveguides for tunable lasers. IEEE Journal of Lightwave Technology 13, 1905–1912; 1995.

[Brü01]   [Brü 01]Brückner, V.: Einführung in die DWDM-Technik. Unterrichtsblätter der Deutschen Telekom Nr. 8, S. 458–469; 2001.

© Springer Fachmedien Wiesbaden GmbH, part of Springer Nature 2024   245
V. Brückner, *Elements of Optical Networking,*
https://doi.org/10.1007/978-3-658-43242-3

[Gla97]   [Gla 97]Glaser, W.: Photonik für Ingenieure. Berlin: Verlag Technik 1997.

[Thi02]   [Thi 02]Thiele, R.: Optische Nachrichtensysteme und Sensornetzwerke. Braunsch-
               weig Wiesbaden: Vieweg & Sohn Verlagsges. 2002.

[Hul96]   [Hul 96]Hultzsch, H. (Hrsg.): Optische Telekommunikationssysteme. Gelsen-
               kirchen: Damm Verlag 1996.

[Hub92]   [Hub 92]Hubmann, H.-P.: Lichtwellenleiterpraxis. München: Franzis-Verlag 1992.

[Mah01]   [Mah 01]Mahlke, G.; Gössing, P.: Lichtwellenleiterkabel. Erlangen and Munich:
               Siemens Publicis MCD Corporate Publishing 2001.

[Opi95]   [Opi 95]Opielka, D.: Optische Nachrichtentechnik. Braunschweig Wiesbaden:
               Vieweg & Sohn Verlagsges. 1995.

[Hei94]   [Hei 94]Heidrich, H. et al.: Monolithically integrated heterodyne receivers based on
               InP, ECOC '94 Firence 1994, vol. 1, S. 77–80.

[Kra02]   [Kra 02]Krauss, O.: DWDM und Optische Netze. Berlin, München: Siemens AG
               2002.

[Elr88]   [Elr 88]Elrefaie, A. F. et al. : Chromatic dispersion limitations in coherent light-
               wave transmission systems. IEEE Journal of Lightwave Technology 6, Nr. 5, 704–
               709; 1988.

[Ebe10]   [Ebe 10]Eberlein D., Glaser W., Kutza, Ch. und Manzke, Ch.: Lichtwellenleiter-
               Technik. Renningen: Expert Verlag 2010

[Gle13]   [Gle 13]Idris S., Osadola T. Glesk I.: Towards self-clocked gated OCDMA receiver,
               Europ. Opt. Soc. Rap. Public. 8, 13013, 2013.)

[Eng14]   [Eng 14]Engelbrecht R.: Nichtlineare Faseroptik: Springer Verlag Berlin Heidel-
               berg 2014.

[Brü15]   [Brü 15]Brückner V.: Lecture Notes in Fiber Optics ; 3rd order nonlinear, nonsta-
               tionary wave equation. ResearchGate 1915

Index

Printed in the United States
Baker & Taylor Publisher Services